# THE ATLANTIC FLYWAY

# THE ATLANTIC FLYWAY

*By Robert Elman*

*Photography by Walter Osborne*

WINCHESTER PRESS

All rights reserved under International and Pan American
Copyright Conventions

Library of Congress Catalog Card Number 72-79367

ISBN 0-87691-083-5

Designed by Alexander Stepanovich

Published by
WINCHESTER PRESS, 460 Park Avenue, New York 10022

Printed in Japan by Toppan Printing Company

*This book is dedicated*
*to Larry Koller,*
*who took delight in sharing with others*
*the sight of distant wildfowl*
*glimmering above the Eastern Shore.*

# CONTENTS

# COLOR PORTFOLIOS

# ACKNOWLEDGMENT

The text of this book mentions major writers and scientific investigators whose works are described, and a selected bibliography lists a number of important authors and their works. In addition, special acknowledgment is made for the unstinting assistance of personnel at the United States Bureau of Sport Fisheries and Wildlife—most especially to Edward Addy, Dan Saults, Dr. Joseph P. Linduska, Marshall L. Stinnett, Dr. Robert Smith, James W. Carroll. Luther C. Goldman, and Philip A. Dumont.

# INTRODUCTION

The scene is the Atlantic Flyway—a mighty funnel of diverse but tenuously connected habitats stretching more than three thousand miles from above Baffin Bay in the Arctic to below the Antilles. The *dramatis personae* comprise a wondrous array of game—more than three score species of ducks, geese, swans, and shorebirds, as well as various furred animals—together with other wild creatures, and man himself. The time of action is the past, the present, and the future which pours inexorably into the past; and the drama concerns the still unfinished saga of man's impact on a former wilderness.

The central theme is a dramatic one, even melodramatic, for although man can hope to correct many of his environmental blunders there remains the constant, terrifying danger that he will commit some truly irreversible folly, some crime against nature which can never be undone. We must hope that our drama is not destined to prove, in the end, a tragedy. We must do far more than hope, of course; we must plan and act wisely. But first, we must care.

Few experiences are as exciting as the revelation of some new secret of the outdoor world. It is hoped that the words and pictures of this book can suggest some of the special kinds of excitement nature herself affords, while at the same time conveying facts which are essential to an informed concern for the Atlantic Flyway and the other three great American flyways with which it interacts. For an understanding of this flyway and others like it is imperative if future generations are to inherit the birthright we have been granted: the wild game and wild beauty of the wetlands.

Many paths lead to nature's exquisite labyrinth. Some may find it strange that my own perception of nature has been most powerful and moving when it occurred within the hunting context, but it is so; in my experience, the activity of hunting provides an access to nature second to none and far richer than most.

A recently published indictment of man's hunting behavior is content to explain it in terms of the seduction of its gadgetry, the masculinity of its social bonding and image, and, most of all, an atavistic fascination with killing. All of these motivations may be significant for some hunters, and surely there is pleasure in the skillful use of fine equipment and the comradery of man and dog, but for most hunters, I believe, the essence of hunting lies elsewhere. True, killing is an element of hunting, as death is an element of all life, and many of us feel the tug of a primeval drive to capture our own food. Yet the skillful manipulation of a gun is almost the least of the wildfowler's accomplishments and only a small part of his hunting activity.

To begin with, he must know how to distinguish between many different species of birds, often at a considerable distance; he must understand how, when, and where they fly, how they feed and on what, how they sound, what they look and listen for. In short, his knowledge of wildfowl must be good enough to fool the birds themselves. He cannot get close to them without getting very close to nature.

Thus the wildfowler's skills, shooting aside, require a considerable knowledge of game biology and behavior, a knowledge demanding careful study and keen perception. Above all, what nature grants to the hunter is a sharpening of interest and comprehension that makes the acquisition of such knowledge possible and brings the hunter back, over and over again, to the marsh or forest or plain.

In addition, my experience with wildfowlers has been that they care more than any other group about wildfowl, and express their concern more vigorously in work and money directly benefiting wildfowl and their habitat. And so, speaking as an evangelist of conservation, I cannot damn my fellow hunters, fashionable though it has become to do so. On the contrary, I trust that they will find much on the following pages that will be of special interest to them, regardless of where or how they hunt.

I certainly hope, however, that non-hunters, ex-hunters, even anti-hunters will also take pleasure and gain instruction from this book. The hunter's access to the mysteries of migration is not a monopoly. Our understanding, our awareness of nature, grows with each day of observation whether it is spent shivering in a duck blind or slogging over the tidal flats as a birdwatcher. Facts, theories, and speculations of equal interest to both groups have been gleaned from the scientists whose profession it is to unlock nature's secrets, and some of their findings are presented in the text. The sources are cited in the hope that they will lead to further reading, but this book does not attempt to be a treatise on conservation, or a history, or, least of all, a description of how to bag ducks. It is, quite simply, an exploration of the wetlands, a plea for their preservation, a tribute to the marshes, the game, and the hunters and non-hunters alike who cherish things untamed.

We need not look back very far to find a time when greater concentrations of canvasbacks gathered over part of the Atlantic Flyway than anywhere else on earth. Yet today this magisterial duck and all other species of wildfowl are menaced by the continuing encroachment of urbanization, industrialization, highway building, drainage by government agencies as well as land speculators, and pollution by virtually everybody. The Atlantic Flyway has withstood all these ravages and others, and if—*but only if*—man can learn to impose upon himself a sane responsibility as an interacting denizen instead of a detached observer of the natural environment, it can continue to survive.

This book, a close collaboration between author, photographer, and designer, was Walter Osborne's idea, the last work this brilliantly gifted writer-photographer was able to complete before his death in early 1972. Walter Osborne was best known for his careful research and breathtaking photography of horses, but he was equally at home in the marsh as in the paddock or the rodeo arena, and the pictures taken for this book were very much a labor of love. A wildfowler in his youth, Osborne preferred to hunt with a camera toward the end of his life. Neither the sweeping panoramas nor the most intimate secrets of the shyest wildfowl eluded his lens, as the portfolios on the following pages reveal. They constitute an invaluable portion of the rich legacy he left on film for all who esteem sensitivity, originality, acuity of perception. We who collaborated with him on this book can only hope that its realization does justice to his conception of it and, in some small measure, to his memory.

ROBERT ELMAN

WISE IS THE WILD DUCK WINGING STRAIGHT TO THEE,
RIVER OF SUMMER! FROM THE COLD ARCTIC SEA,
COMING, LIKE HIS FATHERS FOR CENTURIES, TO SEEK
THE SWEET, SALT PASTURES OF THE FAR CHESAPEAKE…

GEORGE ALFRED TOWNSEND

# THE ATLANTIC FLYWAY

*Restlessly milling over coastal marsh, snow geese prepare for migration.*

# Chapter One

# GENESIS OF THE WETLANDS

In the beginning there was ice. Endless horizons of ice that filled the valleys and passes, reducing mountain chains to insignificance. Rolling prairies of almost lifeless ice, permitting little more than algae to survive, ice abrading the earth and gouging great troughs into the rocky soil. Slowly the glaciers receded as the planet warmed again, leaving buried chunks like icebergs in an earthen sea. And finally the icebergs melted into this solid sea which had begun to ripple with prairie grasses nourished by the sun, toughened by the winds.

Where the chunks of ice subsided into the land, shallow basins remained—the potholes of the northern prairies. Some are mere puddles, and new ones have been forming and drying up with the infinite constancy of change in the millennia since the Ice Age. Others have evolved into shallow, semipermanent marshes, still others into deep, permanent marshes and open bodies of water.

Each spring countless ducks, geese, and other wild creatures seek out these oases in the vast, dry grasslands, for the marshes and sloughs provide miraculous nesting grounds. Because the mating pairs of many waterfowl species require a degree of isolation during courtship, even the smallest ponds help to spread the nesting population. A drought can shrivel these little potholes, and with them the year's reproductive cycle, yet periodic minor droughts are beneficial. The exposure of bottom muds and vegetation releases nutrients so that, upon reflooding, plant and insect life will flourish. The sustenance of new vegetation is enhanced by the transmutation of insect protein into maximum egg production. And the genes of the hardiest birds—those that have survived the dry period—reinforce the hardiness of generations to come.

How different is the self-correcting misfortune of a dry spell from the Dust Bowl droughts inflicted by men who drain wetlands for the sake of such crops as breakfast cereal and highways, electricity and housing complexes, and the poison-spewing monstrosities euphemized as "industrial parks."

Perhaps even more important than the smallest potholes in the wild scheme of life are shallow but larger marshes teeming with insects and emergent vegetation, each offering the shelter and protein of a natural waterfowl hatchery. And on the deeper sloughs, larger potholes, open waters, the young develop to the flight stage, secure from most predators even when the smaller ponds cake beneath the drying sun. Here and along the vast northern tidal flats, ducklings and goslings grow strong while their elders rest and moult, safe from terrestrial enemies.

According to one widely held theory of migration, waterfowl and many other now-migratory birds once led a sedentary existence in the northern latitudes. A relentless glacial thrust prodded them southward in search of food, it is theorized, but they wandered north again during relatively temperate periods. Then, during the long centuries of glacial retreat, each winter forced them south to join the large avian populations of warmer regions. Predation and the competition of other species may explain why they continued to fly north to breed until a migration pattern was established.

There is a reverse theory that most waterfowl were originally tropical or semitropical, and were forced northward to nest when their home range could no longer support a cataclysmic population increase as the planet's climate moderated.

Either theory accounts for the gradual establishment of vaguely longitudinal corridors of migration, indistinctly edged, often overlapping, but somehow engraved like channels of collective memory upon the map of evolution. There are such channels almost everywhere on the globe, some so short as to go almost unnoticed, some thousands of miles long. They follow lines of least resistance and surest survival—coastlines, the courses of great rivers, the slopes of mountain chains, linked valleys, fertile drainages, the most easily traversed distances from one resting and feeding haven to the next, networks of streams snaking southward, lake shores, massive landmarks, and subtle topographic changes which serve, with the stars and sun, as beacons.

*Northern tidal flats are rich grazing meadows for broods of Canadas.*

Much remains to be said—or questioned—regarding the mysteries of migration and the boundaries and very nature of the flyways, but it is at least certain that these immense flowages of life have long insured the survival of the wildfowl multitudes.

It is generally accepted, though with reservations and qualifications, that the North American continent is blanketed by four such flyways: the Atlantic, Mississippi, Central, and Pacific. Each overlaps and blends with the neighboring corridor, and each absorbs stray migrants from distant airlanes. It has therefore been argued that, in reality, there exists just one enormous continent-wide flyway, although some species may be particularly numerous or entirely absent over one or another portion of the great migratory funnel. Perhaps.

But there are those abundances and absences to account for. And there are groups of birds that have followed one of the four alleged routes (notwithstanding occasional deviations of individuals or small flocks) each spring and autumn for many centuries. There are even small, narrow gaps where the four alleged flyways do not quite overlap—or even touch. It would seem, then, that the four flyways must be more than arbitrary geographical slicings imposed by boundary-conscious man. (Though the biologists of the Fish and Wildlife Service sensibly regard the flyways only as administrative divisions, governmental convenience in this case has adopted itself to natural contours.)

Each of the four is vital to the continent's wildfowl. Each is as essential as an article of faith to the men who scull the bays, crawl the reed-rimmed banks of farm ponds, drift in camouflaged canoes and sneakboats, crouch in pits, or huddle in the wind-raked loneliness of elevated blinds.

Yet, in numbers and variety of wildfowl, the Atlantic Flyway is unique among the four. It is the subject of this book for other reasons as well. Geologic and climatic vagaries have pocked its length and breadth with ideal waterfowl havens. In certain respects, it may be the healthiest of the four American flyways—the most promising for the future. It is also the most threatened. The Atlantic seaboard is the most heavily populated, heavily industrialized, heavily polluted, drained, and despoiled zone in the nation. The outcry for correction is loud here, as inspiringly militant as it is occasionally misguided. Progress is being made despite mistakes. The future is a question of whether the corrective measures can overtake and destroy the destroyers. The fate of the Atlantic Flyway may presage the fate of all the wild lands and waters where America's growing population of outdoorsmen seek recreation and solitude.

*Immense sloughs like those of Utah's Bear River Refuge furnish
hunting sites and serve as hatcheries for eastern and western flyways.*

The Atlantic Flyway is, moreover, the birthplace of American wild-fowling for the fundamental reason that it was the area first explored and settled. Its history, ecologically and socially, is an enthralling chronicle of defeat and victory. But before one can comprehend the implications of this history, one must become familiar with the flyway's earth, waters, inhabitants, and skies.

Shortly after World War II, game biologists reported that about fifteen percent of the nation's wildfowl were found on the Atlantic Flyway. In view of the overlap with and tributaries from the Mississippi Flyway, it was an underestimate. It is probable that this cornucopic corridor of nomadic birds accounts for at least a quarter of all the waterfowl of the four great flyways of America. There are nearly sixty species and subspecies of ducks and geese in this country, and more than half of these varieties are classified as legally huntable somewhere along this flyway.

Among the chief reasons for the abundance and diversity of waterfowl along the Atlantic Flyway is the eastern zone's wealth of fecund, if precariously linked, ecosystems.

The prairie potholes are not the sole havens along any of the vast migratory corridors, nor even the exclusive breeding grounds, but such habitat produces at least half of North America's waterfowl—probably far more if one counts areas of mixed habitat that include both potholes and other congenial settings such as tidal flats, river edges, and estuaries. Scattered like a galaxy of life-giving stars, the potholes dot some 300,000 square miles of the continent. And although 60,000 square miles or so are within the United States (chiefly in the Dakotas and western Minnesota), by far the greater part of the hatchery is Canadian. Waterfowl nest as far northwest as Alaska, as far northeast as Labrador. They nest in every province and in the Northwest Territories and in the Yukon. Still, the continent's real waterfowl manufactory is but three provinces wide. It is the immense prairie-pothole zone stretching across southern Manitoba, Saskatchewan, and Alberta. Canada produces eighty percent of the ducks and geese—as well as many of the shorebirds—seen along the four great American flyways, and a disproportionate number of these birds are hatched in the prairie provinces.

One thinks of the prairie sloughs chiefly in connection with the pond and river species of the Atlantic Flyway—"surface feeders" or "dabblers" or "puddle ducks" like the mallard, black duck, gadwall, baldpate, pintail, green-winged teal, shoveler. And it is true that they favor those waters, sharing the habitat with some of the diving species such as the redhead, canvasback, and scaup. But the redhead and canvasback prefer to winter along the estuaries and tidewaters where many other divers are found—the bufflehead, ruddy duck, eiders. And most of the dabblers are not restricted to the prairie range even in breeding season but are also sprinkled along streams and on swamp and bog ponds.

All along the flyways and across the immense breeding grounds, nature has been prodigal in furnishing shelter and foods. The marshes provide smartweeds, pondweeds, waterweed, soft-stem bulrush, rice cutgrass, arrowhead, wild rice, sedges, milfoil, bur-reed, musk grass.

*On pothole waters, well-fed mallard drake fans his wings.*

*"On the Wing," one of Hans Kleiber's watercolors,
depicts rich duck habitat of U.S. interior.*

The damp bordering meadows serve up horsetail, foxtails, rabbitfoot grass, wild millet, and more. In wooded areas, the nut trees, too, spatter the ground with forage for a variety of ducks. Even the bog ponds, whose acid chemistry discourages many plants, are fringed with rosemary, Labrador tea, cottongrass and sedges, cranberries, blueberries, winterberries, bogbeans, chokeberries, wild raisins. The salt marshes provide bulrush, widgeongrass, spikerushes, glassworts, and such. Those plants with little nutritive value have other uses. When hardstem bulrush is available, it is a preferred nesting material of the Canada goose. And cattails and alkali bulrush attract muskrats, whose lodges perform a second function as nesting sites.

The eelgrass, beloved of brant and once so disastrously blighted as to threaten the extinction of the species, flourishes once again, as does the wild celery sought by many birds.

For those ducks that prefer molluscs and fish, the bays and tidal flats and estuaries are equally rich in provender. They can remain so if man halts the carnage of pollution.

*Audubon's ring-necked duck and drake
were painted on Atlantic Flyway.*

*Ring-necked Duck*

*Ontario wildfowlers come ashore with geese at peak of autumn's gunning season.*

In some ways, that great destroyer civilization has actually benefited the continent's wildfowl. On meadows near water, haystacks are nesting sites for geese. Rice, wheat, and corn—growing or scattered by the spendthrift yet economical speed of the mechanical picker—feed countless thousands of ducks and geese.

Essential though the prairie sloughs and potholes are, many wildfowl are flexible of habit or actually prefer far different surroundings. The gaudy wood duck flits about the swamps, a creature of inland pools and streams. So does that unlikely fish-eating tree-nester, the hooded merganser, though the American merganser prefers more open ponds, and the red-breasted variety is more maritime in its pursuits.

The American golden-eye is a diver that behaves like a tree duck; the ring-necked duck, a diver that can plummet to a forty-foot depth to dredge up food, is frequently encountered rising from the shallows of a bog. The Barrow's golden-eye, on the other hand, prefers small mountain lakes, and the eastern harlequin likes the rough water of coastal mountain streams in Newfoundland and Labrador.

If the convoluted, rocky bights of Maine were snapped straight in some unthinkable upheaval, their stretch would equal half the length of the Atlantic coastline. From here down through the coves of Nantucket and Martha's Vineyard, the American scoter thrives, wintering as comfortably as on the milder waters of the North Carolina coast, at the southern tip of its range. The surf scoter and the white-winged scoter share this range and extend it westward, occasionally to the shores of the Great Lakes. All three summer in the remote northwest, undismayed by cold or distance.

Another denizen of the upper eastern coast is the hardy oldsquaw, a species that nests on tundra lakes. And all along the coast are the tidewaters and estuaries—harbors for the bufflehead, eiders, ruddy duck, and redhead.

The coastal bays are also festooned with ragged lines of brant, but other, larger geese flourish all across the continent. Nesting on coastal marshes, tidal flats, inland lake shores, meadows, and fresh marshes, one subspecies or another of the supreme wild goose, the Canada, pours down every flyway.

For many years, game biologists searched futilely for the breeding haunts of snow geese and blue geese. The snows were found nesting along the northern coast of Greenland in 1929. Subsequent discoveries placed snows and blues on Baffin Island and the northern tip of Quebec, around the upper waters of Hudson Bay.

Blues and lesser snow geese have been found in the Arctic westward to Alaska. Flocks of blues and lesser snows mix freely in their travels— a habit ranking as a wildfowl mystery until biologists realized that they were observing not two anserine species but two color phases of a single species. The greater snows winter along the Atlantic coast, but most of the lesser snows and blues funnel down to the Gulf of Mexico, furnishing great wingshooting excitement on the Texas and Louisiana rice fields of the lower Mississippi Flyway. The Mississippi and Atlantic flyways overlap, of course, and there are occasional wildfowl visitors over wetlands far from their normal paths of migration. Furthermore, some of those paths are shifting, shifting so emphatically that wildfowlers of the Atlantic Flyway may one day come to feel quite possessive regarding lesser snows and blues.

Whistling swans share the far northwestern breeding grounds of the lesser snows, along the shores of the Arctic Ocean and the Beaufort Sea. These shy, magnificent birds, which pair for life and, once mated, never separate, wing south and eastward in the autumn, to bless both the Atlantic and Pacific coasts with their fierce beauty. The mute swan, a graceful European immigrant, has reverted to the wild and established breeding areas on Long Island and in coastal enclaves from Massachusetts down through the New Jersey seaboard to the region

*Food-rich shallows welcome mallards wintering in North Carolina.*

of the Delaware and Chesapeake. Both species can be seen drifting on fresh or brackish shallows, or sailing above the coastline of the Atlantic Flyway.

The riches of this migratory corridor also include a variety of shore and wading birds—plovers, sandpipers, dowitchers, turnstones, killdeer, godwits, curlew, herons, egrets, bitterns, ibises, cranes, rails, gallinules, true coots (webless-footed birds unrelated to the scoters that are "sea coot" to a Down East gunner), oystercatchers, avocets, tattlers, whimbrel, yellowlegs, dunlin, sanderlings, and phalaropes.

And how is one to classify the jacksnipe, or Wilson's snipe? And the woodcock? Both may be called sandpipers, shorebirds misplaced or detoured by evolution, though the snipe has remained a wader of sorts on the margins of bogs, streams and marshes, while the woodcock has ventured farther inland and upland. The woodcock's migration is a subject of fascination—the sly, sporadic, nocturnal flights of early autumn by which large concentrations appear and disappear unobserved, the massing at Cape May for the fearsome, seemingly suicidal flight across Delaware Bay, the sudden, never quite expected appearance in Louisiana woodlands.

Yet it is the multitude of waterfowl that most beguiles the observer along the Atlantic Flyway, birds that sometimes drop innocently before the blinds or decoy to a bobbing jug, and are at other times as elusive as mist blowing off a far shore, birds that sweep out of the northern skies in veils, apexes, and columns of majesty. The full appreciation of these mysterious birds, the love and respect with which a man accepts a breakfast-size teal or the sagging weight of a Canada from his steaming retriever, is a privilege of those who have seen the dawn glow upon the wetlands.

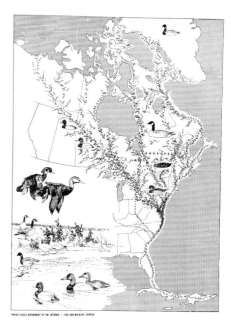

# Chapter Two

*Blending, interlocking nature of
America's major corridors of wildfowl
migration can be seen on maps
prepared by Fish and Wildlife Service.
From left: Atlantic, Mississippi,
Central, and Pacific flyways.*

# SOUTHWARD BOUND,
# AND EASTWARD

If we can trust the maps of the United States Fish and Wildlife Service, four major flyways serve the migratory needs and mysterious whims of North American wildfowl.

Each of these corridors is more or less funnel-shaped. The widest at the top, spanning the greatest expanse of northern breeding grounds, are the Atlantic and Mississippi flyways, which are probably also largest in terms of total area covered. It may be that the Atlantic is both wider in the arctic regions and larger overall than the Mississippi, but one cannot be dogmatic in these matters because tremendous areas of the breeding region produce wildfowl for both flyways—in fact, for the Central and Pacific flyways, as well. Moreover, the migratory routes of many birds begin on the Mississippi Flyway and then, for reasons not fully understood, wheel eastward, to terminate on the Atlantic Flyway.

The blending, interlocking nature of the corridors, and the tributary lanes crossing from one to another, make it difficult to assign precise eastern and western limits to the courses of the aerial caravans. Though the Atlantic Flyway is bounded on the east by a vast oceanic expanse, the seaboard is sometimes visited by the European barnacle

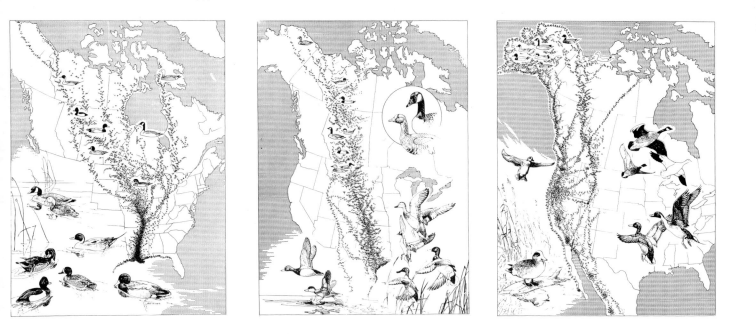

goose, which breeds on the Norwegian island of Svalbard, northeast of its other primary breeding site on Greenland, and is extremely common in winter along the west coast of Scotland. Even its name is purely European, having derived from an old Norse belief that the species hatched out of barnacle shells on water-soaked logs and driftwood.

The European widgeon also visits the Atlantic Flyway, from New England to Florida, and sometimes ventures far inland. It breeds in Iceland, and may be more common in North America than its name implies. A third visitor from the Old World is the European teal, a close relative of our greenwing. Since teal seem to have a propensity for incredibly long migratory journeys, it is perhaps not surprising that this "European" species also appears throughout Asia, breeds in the Aleutians, and visits North America's eastern seaboard from Greenland to North Carolina.

There are also migrants from the far Northwest, birds that touch all of the major flight corridors on their way to winter quarters. The Atlantic Coast receives some canvasbacks and scaup, for example, from Alaskan breeding grounds.

Still, if one bears in mind that the boundaries fade gradually and that migration patterns occasionally shift, it is possible to define the flyways geographically. The Atlantic Flyway gathers hordes of wildfowl from southern Greenland, the islands of Baffin Bay, Hudson Bay, Labrador, Quebec, the Gulf of St. Lawrence, and all the Maritime Provinces, Ontario, most of Manitoba, northern Saskatchewan, the northeastern corner of Alberta, and most of the Northwest Territories up to the northernmost segment of the Yukon border. The prairie prov-

11

inces and eastern Canada are tremendously productive, but many of the birds breed as far south as New England and New York, Michigan, Wisconsin, and northeastern Minnesota.

Some species nest above or below this primary breeding range. The islands of the eastern Arctic are important hatcheries for Atlantic brant and snow geese. Mallards and black ducks are common nesters from Maryland northward. Wood ducks breed throughout the flyway.

The western limit of the flyway angles through the northeastern corners of Minnesota and Wisconsin, a large portion of northeastern Michigan, Ohio, and West Virginia, then turns sharply southward through Virginia, the Carolinas, Georgia, and Florida before sweeping southeastward again to the southernmost wintering grounds in the West Indies. Perhaps one cannot·even classify the West Indies as the southernmost wintering grounds, because a few birds do fly on into South America. Since the ocean forms the corridor's eastern boundary, the Atlantic Flyway also encompasses Maine, New Hampshire, Vermont, Massachusetts, Connecticut, New York, New Jersey, Pennsylvania, Delaware, and Maryland.

For the Mississippi Flyway, the breeding grounds extend from Baffin Island and western Quebec westward through Ontario and the greater portion of the prairie provinces, all of the Northwest Territories, the northern Yukon and northeastern Alaska. The flyway proper covers most of the Midwest, from the amorphous edge of the Atlantic Flyway to the eastern Dakotas and Nebraska, and down through Arkansas and Louisiana to the shores of the Gulf. The upper part of this funnel heavily overlaps the Atlantic Flyway, but there are thinning stretches and even narrow gaps between the two corridors in eastern Kentucky, Tennessee, and Alabama, and in Florida's western panhandle where the shooting is generally not quite as good as it is slightly to the east and west.

The Central Flyway overlaps the Mississippi Flyway and continues much farther southward—through Mexico and all of Central America. It is fed not only by the prairie provinces (chiefly Saskatchewan and Alberta) but by most of the Northwest Territories and the Yukon, as well as northeastern Alaska. It should be noted that the breeding grounds for both the Mississippi and Central flyways are by no means confined to the regions above the forty-ninth parallel. Among the major nesting areas of the redhead, for example, are large prairie-slough tracts in Minnesota, both Dakotas, Nebraska, Montana, Idaho, Wash-

ORIGINS
AND
DEPARTURES

Flight of brant weaves northward in spring. As summer
wanes, sprig's eclipse plumage will change to autumn
finery and young wood duck will fly. Mallards, courting
(below right) and feeding in company with hybrid, will have
raised their brood, and ice will form where black ducks forage.

Despite untidy moult and spatulate
bill, shoveler drake retains slight
dignity, but lobe-toed American coot
is awkwardly comic. Amusing, too, is
pursuit of rival by mallard drake. Farmland
and potholes nurture both geese and ducks.

From northern tidal flats like this lush nesting site, Canadas reach every flyway. Mute swans brought from Europe now breed from Chesapeake Bay up into Massachusetts.

*Canada geese hurriedly take off after rest stop during southward journey.*

ington, the region where California, Oregon, and Nevada meet, and (historically one of the continent's greatest nesting sites) the Bear River Marshes at the northern end of Utah's Great Salt Lake.

A great many redhead ducklings have been banded at the Bear River Marshes, and their remarkable migration routes illustrate why a discussion of any flyway inevitably involves other flyways. The autumn migration of redheads from the prairie provinces is divided into routes heading west, southwest, south, and east-southeast, so that some of these birds show up on every flyway. But a few of those from the Bear River Marshes follow a much stranger autumnal route. They begin by flying *north* through Idaho and Wyoming, then turn eastward across Montana, the Dakotas, Minnesota, Wisconsin, and Michigan. About here, they begin joining the prairie-province redheads and fly on into the Finger Lakes region of New York or to the Maryland and Virginia coasts. Having begun as residents of the Central Flyway, they have crossed the Mississippi Flyway and become Atlantic Flyway birds.

Migration routes and corridor boundaries are for the most part determined by a combination of resting and feeding areas along the way and by the geographic lines of rivers and mountains—the Missouri and the Mississippi, for example, and the Alleghenies which edge the Atlantic Flyway. No satisfactory explanation has yet been promulgated for the manner in which some Bear River redheads defy the rules. One may speculate on ancestral migration routes, established prior to geographic and geologic changes that now make those routes seem peculiar. But migration patterns—and wintering areas—are shifting even now.

For example, there is the cattle egret (*Bubulcus ibis*), the only Old World bird to "invade" the United States successfully without man's invitation. It is not merely a casual visitor like the barnacle goose and the European teal and widgeon; it has become a permanent resident. It is now common and is spreading its range across the eastern half of the country.

Then there is the sudden buildup of avocets at Delaware's Bombay Hook Refuge. Perhaps this can be explained by a general increase of avocets, and the concomitant expansion of their range, during recent years, but other factors may also be at work. A substantial increase

*Redhead dives for food in deep waters of coastal bay.*

in the glossy ibis population has been witnessed at New York's Jamaica Bay Refuge, and this may signify a healthy or unhealthy change. That is, it may be the result of a general healthy expansion of numbers and range, or it may be caused by man's canal-building, swamp-draining, Everglades-shrinking, habitat-wrecking follies in Florida. Wild creatures are of necessity opportunists. It is hardly surprising that they are as much affected by changes in land and water use as by management programs.

To cite a final example, the snow geese of the Atlantic Flyway have traditionally been *greater* snows, but this, too, is changing. As noted earlier, the supposedly western lesser snows and blues, for reasons not yet understood, are shifting their migration patterns. More and more of them are appearing along the middle and lower reaches of the Atlantic coast.

To return to the flyway perimeters, the western edge of the Central corridor runs through western Montana, eastern Idaho, the northeastern two-thirds of Utah, most of New Mexico, and onward to points south. It straddles the Rockies, rather than being hemmed by them, for while most birds follow the eastern slopes in their migration, some parallel the western slopes.

The Pacific Flyway is the narrowest of the four since it extends only from the Rockies to the ocean, but it is also the longest. It draws migrants from all of Alaska and the Yukon, all of British Columbia, most of the Northwest Territories, and Alberta. Like the other flyways, it also draws birds from nesting areas within the United States—some of those Utah redheads for example. Then there are the several strains of western Canada geese, whose breeding grounds stretch from the Arctic as far south as Wyoming. Baldpates nest from Alaska as far south as Colorado; buffleheads from Alaska to California; canvasbacks from Alaska to Utah; gadwalls from Alberta to Nevada; American golden-eyes from Alaska to North Dakota; mallards from Alaska and most of Canada to California and eastward through Nevada, Utah, Colorado, Nebraska, Iowa, Wisconsin, Ohio, clear into New York; mergansers of one kind or another from Alaska through most of Canada and most of the United States; pintails from Alaska and western Canada down to Colorado; ringnecks from the prairie provinces to Wisconsin; ruddy ducks from the prairie provinces to Nebraska; lesser scaup from Alaska to Nebraska; white-winged scoter from Alaska to North Dakota; shovelers from Alaska to Colorado and many points east and west; blue-winged teal from the Yukon and

Northwest Territories down to Colorado, west to California and east to New York; greenwings from Alaska down into California and east to Iowa.

That list does not include species that are exclusive to the west or east but only those that have significant breeding grounds within the contiguous United States, and cross two or more flyways, and are found on the Atlantic Flyway as well as along some or all sections of the Pacific.

The Pacific Flyway reaches all the way down into South America, thanks in small part to that astounding flyer, the tiny blue-winged teal, which, ironically, turns inland and is rarely encountered on the West Coast of the United States. Because bluewings prefer the interior, the majority of them use the Central and Mississippi flyways for a great part of their journey. But they are, nevertheless, common on the Atlantic Flyway, and some of them breed around the Great Lakes and in northern New England.

They are the earliest of migrants, southward bound by the time of September's full moon, and many of the older males have gone on ahead in August. These birds make longer migratory flights than any other North American duck. One, banded in Saskatchewan, was retrieved six months later in Peru, more than seven thousand miles away.

Frederick C. Lincoln, the great authority on migration who carried out pioneering research for the Fish and Wildlife Service, recorded a blue-winged teal that was banded in Quebec on September 5, 1930, and was killed on the second day of October in British Guiana—a distance (as the teal probably flew) of some 2,400 miles.

When our indolent mallards arrive at resting and feeding waters during migration, they may remain there until the next night or they may linger for a month, but this teal flew at least eighty-five miles a day for twenty-eight days.

If that was a record at the time, it has certainly not remained so during the decades of large-scale banding and study. In 1969, the *Portland* (Maine) *Press-Herald* carried an item by Gene Letourneau reporting that a teal banded at Maine's famous Merrymeeting Bay on September 10th had been shot twenty-four days later in the West Indies, some three thousand miles away. The bird had covered an average distance of one hundred and twenty-five miles a day.

One might almost fancy that its gypsy call as it rushed along was a harried, White Rabbitish self-scolding: "I'm late! I'm late! I'm late, I'm late, I'm late . . ."

GREEN AND BLUE WINGED TEAL.

*Teal on ponds were popular subjects in 19th-century art.*

Bluewings have an impressively wide breeding and wintering range. Though many of them fly as far south as Chile and Brazil, others winter in the Gulf states and along the lower reaches of the Atlantic Flyway—principally in Florida, Georgia, and South Carolina.

Frequently a special hunting season is opened early and exclusively for teal. Gunners of the East have long been intrigued and often frustrated by these lovely little cobalt-coverted aerial demons, so easy to decoy, so difficult to shoot.

They, too, have exhibited a breeding and migratory shift, but in their case no mystery is involved. Since 1900, over half of their American breeding grounds have been drained and put to the plow—6,000,000 acres in Iowa alone. And in New England and eastern Canada, towns have sprouted, lands have been "improved" and industralized. Little wonder that the major nesting areas are located inland and northward. The three prairie provinces now produce about eighty percent of the birds. Not quite as many are seen on the upper Atlantic Flyway as were present when Thoreau wrote of wildfowlers "going down through the meadows with long ducking-guns, with water-tight boots wading through the fowl-meadow grass, on bleak, wintry, distant shores . . . and they shall see teal—blue-winged, green-winged—sheldrakes, whistlers, black ducks . . . and many other wild and noble sights . . . such as they who sit in parlors never dream of."

In 1840, Audubon recorded his observations of the species, emphasizing its remarkable aerial abilities: "The flight of the Blue-winged Teal is extremely rapid and well sustained. Indeed, I have thought that, when traveling, it passes through the air with a speed equal to that of the passenger pigeon."

17

*Mallard rises en route from coastal stream to cornfield.*

Audubon was wrong, of course. Most of the diving species can surpass teal in sustained flight, and even mallards have been observed to outdistance them when a mixed flock was pursued by a peregrine. Teal are fast, but their small size makes them appear to be going even faster. The illusion is enhanced by sudden twistings and turnings—perhaps even more characteristic of green-winged teal than of bluewings. They can perform these maneuvers in unison, and some ornithologists have speculated that they may actually have an unknown method of communication to prevent aerial collision. Although careful observations have proved that heavier birds can achieve greater speed in sustained flight, startled teal can spring from the water as vertically and swiftly as a partridge flushed from concealment. This ability, too, has embroidered the legends of speed.

The green-winged teal does not fly as far south as the bluewing, the lower terminus of its winter range lying across Mexico, but it, too, frequents the Atlantic Flyway. Significant numbers of greenwings winter in Atlantic states, as far north as Virginia, and some of them pass through the upper flyway en route to their destination.

Certain migratory mysteries are evident in connection with these immense diversities of winter range. It is known that birds of a single species from a single nesting slough may not all fly to the same winter quarters, nor will all of them follow the same migratory route. They may actually head for several different flyways. A recent report, based on a seventeen-year Illinois banding study, shows that mallards and black ducks will choose several directions from a large nesting or resting pond. Most of the Illinois ducks follow the Mississippi Flyway to Arkansas, Tennessee, Mississippi, and Louisiana, but many turn eastward when they reach the lower latitudes—and many others cross over to the Atlantic Flyway at an earlier stage of migration.

The average count of wintering mallards each year from 1960 to 1966 was 113,000 in South Carolina, 5,000 in Georgia, and 3,700 in Florida. Almost all of these had come from the Midwest and crossed over from the Mississippi Flyway. (Since more mallards are found to the north, and their winter quarters stretch from coast to coast, the South Carolina figure is only a tiny indication of the incredible population of the ubiquitous mallard.)

An immature mallard or black duck will ally itself with adult birds and take off with them. Their route and destination seems to fix the route and destination of the immature bird—not only that year but in years to come. The first journey seems to imprint itself like a map in the bird's migratory instinct and the duck will become, on subsequent migrations, ever more proficient at homing to resting places and the final destination.

Another immature duck from the very same brood may choose to ally itself with different adults, leaving on a different route for a different destination—not necessarily even on the same flyway. Nature seems to spread the population out quite satisfactorily—but how? Is the immature duck's choice of migratory mentors based purely on chance? And by what mysterious process do the adults form migratory flocks at the start of the journey? How does each individual, imprinted with a particular route and destination, know which other birds of the same species on the pond are similarly imprinted? And why, on infrequent occasions, will a bird or group of birds abruptly defy that imprinting, either switching routes or remaining in the comfortable South and abandoning migration forevermore?

Much remains to be learned, and some of the findings will be of great value in game management and hunting regulation. Because Canada geese migrate in families or groups of families, it is fairly easy to regulate their harvest by single migration lanes within flyways. But predictions of abundance on a single lane are more difficult for a few of the duck species. Current studies may shed new light on their flight patterns and shifts.

Still another mystery is the wonderful homing ability of wildfowl that have made that first, imprinting migratory journey. The majority of them will subsequently home to within sixty miles of a banding station, which does not seem terribly impressive until study reveals that the sixty-mile distance is a *foraging* radius and the birds are capable of returning consistently to the very same pond from which they departed. Banding experiments have proved wood ducks, black ducks, and mallards to be very accurate homers—and the older they are, the more accurate they are. Buffleheads may be even more adept at homing. In one study, twenty bufflehead hens were banded, and eighteen of them returned to the same water. The other two came back to nearby lakes. Two of the eighteen were found using the same nests for three consecutive years.

It has also been found that ducks are conservative by nature, They will home to an established area even in spite of drought, both in the course of migration and upon reaching winter latitudes. Then, if the habitat is not satisfactory, they will begin exploring for other waters, and sometimes a flock, finding other birds already using a pond, will leapfrog to a more distant haven. But as a rule—and with incompletely understood exceptions such as the shift of blues and snows—changes in the migration pattern are so gradual as to be frequently imperceptible.

Ancient shifts of this sort may explain the curious division of routes among the Canada geese and black ducks of the Atlantic Flyway. The New England gunners who defy the frigid winds, patiently watching the northern skies for Canadas and blacks, are harvesting birds from Labrador, the Maritime Provinces, Newfoundland, and New England itself. Few of these Canadas and blacks come from the interior.

Yet the majority of the black ducks and almost all of the Canada geese bagged in the South or on the middle-Atlantic coast do come from the interior. The black ducks come chiefly from Quebec and Ontario. Their route is joined, somewhere along the southern Ontario peninsula, by the imperious wild geese from the eastern shores of Hudson Bay. Some of the Canadas and blacks follow the Mississippi Flyway. Most, however, swerve over Pennsylvania and West Virginia to the Chesapeake and Delaware bays. Some will winter there, some will continue southward.

Along the way other species have appeared, for the same airlane from the northwest is taken by many wildfowl, and particularly the diving ducks. All the divers of the wintry coastlines come from the interior. The gunners of Back Bay, Chesapeake Bay, the renowned Eastern Shore—in fact, most of the great Delmarva Peninsula—occupy boats and blinds just at the gaping mouth of this immense migratory cornucopia. Along the same route have come almost all of the area's scaup, redheads, and canvasbacks.

For that matter, the huge rafts of greater scaup on Long Island waters and the scoters on New England's coast also come chiefly from the interior, although on slightly different routes.

An article contributed by the migratory expert Frederick C. Lincoln to F. H. Kortright's scholarly book, *The Ducks, Geese and Swans of North America,* points out that the Atlantic Flyway has at least three primary routes of migration and as many more tributaries of great importance. "Many of these birds," Lincoln says of the migrants, "are actually flying at right angles to others of their own kind that are following routes from the north to the Mississippi and Central flyways." He notes that an odd tributary is a branch from the Mississippi Flyway which abruptly leaves the river of flight—for reasons not yet clearly understood—near St. Louis, to strike out across the mountains for the South Carolina coast. Many teal, some gadwalls and shovelers, and perhaps some ringnecks, choose this curious, meandering migratory stream.

The Atlantic Flyway is, indeed, as Lincoln succinctly described it, "a complicated system of migration routes."

*Pintail is among species that frequent several flyways.*

# Chapter Three

# PORTRAIT OF A SWEET-WATER MARSH

With the moon beyond the last quarter and the sun not yet up, a grove of scrub pines is barely discernible above the marsh. A gunner trudges along gingerly, intent on getting his decoys set out before the legal shooting hour. He hears only the crunch of sand underfoot as he passes among the pines, then the gurgling of running water and the suck of ankle-deep mud in a narrow cedar swamp. He is mildly startled by the rustle of cordgrass whipping against his waders as he descends a meadow. Thinking about the morning to come, remembering other mornings, he is hardly aware of the sound his dog makes, padding, panting, splashing through each puddle. Several meadowlarks dart chest-high across the gunner's path, close enough to have touched but visible for only an instant as a black-on-black flickering. He smiles, reminded of the hidden life surrounding him.

This, in part, is why he has come here. This, and the difficult beauty of patience in a cold wind, of enticing ducks and geese that have seen decoys before and survived other calls; this, and the smooth gun swing, the report, the plummeting bird, the eagerness of the big black Labrador retriever. It is atavism, no doubt, but also the craftsman's pride and the naturalist's curiosity.

The gunner pauses at the shoreline of the slough, listening to the rhythmic lap of water and the whisper of feather grass, and another whisper contending with the breeze. He has heard no wildfowl cry, but he recognizes the sibilant throb of wings. Then he hears the voice of a flight, a reedy stutter of quacks followed by a staccato rattling. Black ducks, he guesses. It sounds like their feeding call, yet the tone seems both urgent and plaintive. The weather has been cold to the north. Perhaps, despite the waning moon, the flock is setting down after a long migratory flight. Ducks and geese often travel far at night.

Before the morning is over, the gunner hopes to see mallards, pintails, and Canada geese as well. Literally dozens of waterfowl species rest and feed at this marsh on the Atlantic Flyway. Many can be legally hunted, though the regulations and bag limits often vary from fall to fall, based on hunting pressure and a careful watch of the northern breeding grounds. Game biologists have studied water and food levels through the spring and summer, checked hatching and fledging success, taken a census of the anatidine population, closely observed the fits and starts of onsetting migration. The gunner carries a copy of his state's regulations—which, of course, conform to applicable federal regulations—but he will probably have no need to consult it. He knows this year's bag limits, and he can recognize the species along his flyway.

In the pale glow that has now inched upward from the horizon he sees a wavering line of dots curve across the sky toward him. A few of the dots pass the others and a few slip back, forming a wide bow that quickly stretches and sharpens to become a massive chevron. The black silhouettes are Canadas. Now he can hear the resonant calls of the wild geese, bugling cries above the dim, echolike chorus of distant stragglers.

Clambering down the bank to where his boat is tied, he throws back the protective tarp, carefully props his long 12-gauge double against the stern seat, sets his shell box next to it, and stiffly climbs in. His side-by-side is prized for its efficiency as well as its traditional beauty of contour. Only in recent years have such fowling pieces been overshadowed by over/unders and repeaters—the latter equipped with magazine plugs limiting their capacity to three shells in compliance with a sensible federal law. But since few men are quick enough to get off three effective shots before a flight is gone, the choice of gun is hardly more than a matter of personal preference.

The gunner on this marsh has used the same double for years, and he lays it in the boat with a tenderness not usually displayed in handling mechanical contrivances. His big Lab has jumped aboard and clambers among the decoys, whining impatiently.

Keeping near the shoreline, the gunner poles quickly to his blind and then works quickly but methodically to set out his decoys. He starts with several strings of two or three blocks each, lined together and anchored so that they will bob on the current with a lifelike movement. He scatters other blocks about in what appears to be a random positioning. His father, an old Eastern Shoreman, would have set them in a careful "fish hook" pattern, but the value of those traditional arrangements was never more than myth. The gunner follows simpler rules, separating each decoy by several feet from all the others to ensure that all can be seen readily and recognized from the air. In the midst of the spread is a vacant area—landing space. The innermost decoys are a bit over twenty yards downwind from his blind, so that the birds will pass before him as they set down into the wind.

*The late Larry Koller, notable wingshot and one of America's most evocative hunting writers, glows with contentment as he returns from marsh blind in Maryland's farming country.*

The blind is built on the lee side of a great curving spit, overgrown with tall grasses. On the far side of the spit is a river, a second shore-line, and a sprawling farm. Reeds and sedges line the shores, with meadow foxtail growing beyond. Farther off, on the farm, are potholes adorned with sago and rimmed with Japanese millet. Best of all, there are the fields of corn that have drawn gigantic concentrations of water-fowl to the middle portion of the flyway in recent years.

As the gunner places his decoys, he reflects on two recent days of rather mediocre shooting. The birds had been foraging nocturnally under a full, bright moon, and there was little movement after sunup. Then, too, there had been calm, bright high-flying weather on the first day, and on the second a variable breeze that frequently brought ducks over with the wind at their tails. But today, he notes, the wind will be at his back, and a heavy overcast will keep both ducks and geese flying low, watching for landmarks and staying under a layer of turbulent weather. Different conditions may be advantageous in some spots, depending on local terrain and type of habitat, but this is the kind of day when many blinds are occupied on the marshes of the Atlantic Flyway.

The gunner is using two dozen mallard blocks and a pair of Canada goose decoys. If he had a partner with him, he would probably set out more, but even a dozen mallards will usually entice the puddlers. On one of the big coastal salt marshes, or a bay, or on good deep wild-celery water, he would substitute several dozen scaup or canvasback decoys, for these will lure the diver species of the flyway. It takes a big raft of decoys to interest the flights of divers on the open waters of the coast.

A wildfowler of long experience, he sets out a pair of ''confidence'' decoys—oversized Canada goose blocks representing two old traditions of the Atlantic Flyway. Decoys carved in feeding or preening position, or with their heads turned back upon their bodies in an attitude of sleep are sometimes called confidence decoys, the theory being that their serenity proclaims all is well. But a true confidence decoy imitates a species not being hunted, particularly a bird known for wariness and intelligence, and therefore calculated to impart a sense of security to any wildfowl flying over to inspect the rig. Confidence decoys have been in use along the flyway since the mid-nineteenth century, and their use has spread westward. They have included geese, shorebirds, gulls, herons, terns, and even loon in upper New England. Occasionally one still sees a gull—the creation of a wood sculptor or a taxidermist—perched upon the camouflaged bow of a sneakboat.

The other tradition, that of mammoth decoys, is thought to have originated with the market shooters and guides at the famous "shooting stands" erected at Cape Cod during the nineteenth century. These men used live decoys near the blinds, and farther out they floated huge, barrel-bodied "slat" decoys to attract flights from great distances. The idea soon spread southward and appeared in the form of solid blocks as well as slat decoys. A number of them can be seen at the Shelburne Museum in Shelburne, Vermont, where the collections include that of the late Joel D. Barber whose 1932 book *Wild Fowl Decoys* was the first treatise on the history and use of American decoys.

The oversized Shelburne decoys include a primitive solid wood black duck made by Wilbur Corwin at Bellport, Long Island, in about 1850; a solid canvasback made by Wilton Walker of Tulls Creek, North Carolina, in about 1900 and used at Knotts Island, on Currituck Sound; another oversized can, of unknown date, by the renowned Shang Wheeler; and a massive slat goose by the great Cape Cod maker Joseph C. Lincoln. Probably built at about the turn of the century, the Lincoln decoy is over five feet long and bears a passing resemblance to an upturned boat. The use of mammoths spread to other flyways, and such well-known commercial manufacturers as Mason's Decoy Factory of Detroit produced them in the late nineteenth and early twentieth centuries.

Oversized canvasbacks and redheads were popular among market gunners from Back Bay, Virginia, down through Currituck Sound in North Carolina. Even more imposing than the ducks or the giant Canada blocks were the swan decoys used from Currituck north into Chesapeake Bay until swan shooting was prohibited in 1913. Professional gunners sometimes included a few swans in their rigs, and when that law was passed, Joel Barber recalled, many men knocked the long necks and heads off the blocks and converted the decoys into geese. Some of those swans were truly works of art. A specimen from the Barber-Shelburne collection, made in about 1890 by Samuel T. Barnes of Havre de Grace, Maryland, is among the finest examples of tidewater sculpture.

Fifteen-pound wooden swans are no longer encountered on the coastal marshes, but somewhat oversized Canadas are still occasionally seen. Bobbing ponderously on the waters of today's Atlantic Flyway are plastic mallards and even blacks over two feet long and nearly a foot wide, about a third larger than the more common standard decoys. The manufacturers sometimes advertise these looming figurines as ''Magnums,'' a term as apropos of decoys as of shotgun shells or champagne bottles.

If the gunner were out primarily for Canadas, he would use as many goose decoys as possible, whether on water or at a feeding field. It is the nature of these birds to be skeptical, and reassurance is most effective in the form of large numbers of their own kind. Just a pair of them, used as confidence decoys, may or may not draw any Canadas within range. Probably they will attract a few stray singles or pairs. And they

*Mallards glean shoreline and (right) gunner rises for high crossing shot.*

will reassure some of the less wary ducks such as mallards. One of them is a floating block, and the gunner places it at a dignified distance from his duck decoys and very close to shore. The second is a "stick-up," with a pointed stake protruding from its base. He drives this into the mud, standing the decoy on shore near its mate.

It is almost sunup when the last decoy is set. The legal shooting hour has arrived. The sound of chuckling, gabbling calls has become closer and more frequent. The gunner poles his boat over to the blind, an elevated affair constructed on stilts a little out from shore. He had originally planned to build his blind on the bank, where tall grasses would add to its camouflage, but he had found no way to conceal his boat. With the blind on the water, he can slide the craft beneath the elevated floor. The dog jumps out and sits in a corner of the blind, trembling slightly. The gunner opens his shell box, loads up, and settles himself comfortably on a battered piano stool. He thinks about shifting a couple of the nearest decoys. There is no need for the boat since he is wearing his waders.

As he is about to stand, he hears a sudden loud chorus of quacking. A score of mallards rises out of the tall grass sixty yards behind the blind and wheels away. He grins, realizing that the ducks have been quietly resting or feeding there since before his arrival. His activity has finally alarmed them, but he has not actually flushed them and he has not fired a shot. They will probably return later, a few at a time. They will circle his rig, and he will be ready.

A small bunch of sprig passes, fast and too high, their wings flailing so rapidly as to appear blurred. He hopes there will be more pintails— more and closer. A trio of tiny greenwings darts by, dipping and turning erratically. The teal season has already closed for the year and he does not raise his gun.

Beginning to feel chilled and dejected, knowing that the best period for shooting is passing too quietly, he sips hot coffee and remembers other mornings. For a moment he thinks the call of a single wild goose is part of his reverie, but then he sees the big, lone Canada, wings thrusting deeply, neck curved slightly downward, flying low, searching. The bird has seen his confidence decoys and will pass just close enough. The gunner is on his feet, swinging the muzzle past the great gray body before a blast from the right barrel reverberates over the water.

In a few moments the dog is back in the blind, dripping, tail thumping. The goose lies in an opposite corner of the blind.

At mid-morning, the gunner stands and stretches. The goose is flanked by two pintails now, both drakes, one with its iridescent brown

*Silhouetted above gaunt, leafless trees, geese trade over fresh-water marsh.*

head and slender white neck draped across the brownish-gray wing of the Canada. The long tail feathers of the second sprig are cocked upward from the floor of the blind, peculiarly jaunty even when lifeless.

The gunner has not filled his limit, but with the day growing warm and still, he has half decided to quit when he hears the low, chattering feeding call of mallards. He crouches, gun ready, and answers their call with his—tolling them in, not overdoing it, coaxing. Five birds come over, setting their wings to pitch in among the decoys. His gun swings and booms, hesitates an instant and booms again. A hen plummets into the water directly in front of the blind. A large drake arcs downward and crashes on shore with a thudding sound. A double. He is still one bird short of this year's daily bag limit in his state, but he does not care.

As his Lab fetches the drake—the farthest bird first—he is already putting his gear back into the boat. The sky is still overcast and gusts are again rippling the water. For a wildfowler on the Atlantic Flyway, it has been a beautiful day.

# COASTAL
# INTERLUDES

In scenes of arrival, foraging, and rest, snow geese wing over Pea Island, North Carolina; Canadas explore Santee, South Carolina; mallards cross Maryland fields; and canvasbacks raft up on Silver Lake, Delaware.

*Crowded by ducks as well as by their own kind, Canada
geese mill about Connecticut pond. Woodie drakes are
chattering as they leave shoreline of Georgia slough. Bufflehead
crosses mouth of New Hampshire tidal creek. Feeding brant follow
tides along their famous Brigantine, New Jersey, salt marshes.*

Waters may be flat as glass or corduroyed by breezes, and hunters pray for ducking weather. Teal on tiny islet are early-season "breakfast ducks," but bluebills herald wintry months.

At rest in shallows, widgeon take little notice of plump dowitcher.
Redhead couple is expected flyway sight, but Bahama pintail is rare visitor.

# Chapter Four

# FURTHER MYSTERIES OF MIGRATION

The many factors of evolution—random mutation, natural selection, adaptation, survival of the fittest—have resulted in many wonderfully specialized creatures, but none more so than migratory birds. They are marvels of efficiency. It can be argued that the most efficient of all American species is the golden plover, a shorebird of the family *Charadriiae* whose annual journeys outdistance even those of teal.

This plover travels about eight thousand miles on its migration, chiefly using the Atlantic Flyway on its southward journey, though it returns by way of the Mississippi Flyway. The distance is known to be surpassed by only one other avian creature, the Arctic tern, which sometimes breeds within ten degrees of the North Pole and courses down the west coast of Europe and Africa to winter in the Antarctic.

The plover's autumnal flight is mystifyingly circuitous. Upon leaving the arctic breeding grounds in late summer, it wends eastward to Labrador and follows the coastline down to Nova Scotia. But there it leaves the coast to strike out on a hazardous journey over the open ocean, not touching land again until it reaches South America. No one has yet discovered why this astoundingly difficult route has been established or why, in spring, the golden plover returns from Brazil or Argentina along an inland corridor, eventually flying northward along the Mississippi Valley.

Though the reasons for such migrations remain a mystery, biologists now understand some of the adaptations that make possible incredibly long flights. The metabolism of the plover, responding to seasonal hormonal changes, adjusts so exquisitely to the challenge of the long journey without food, without rest, that its forty-eight-hour nonstop oceanic flight of twenty-four hundred miles is fueled by two ounces of reserve body fat.

The golden plover is by no means the only species possessed of such metabolic efficiency or motivational caprice. Many ducks have been sighted flying over the Atlantic, two hundred miles offshore. Some of these birds are bound for Bermuda, the Bahamas, the Antilles. Bermuda does not offer much in the way of winter habitat, yet there have been casual island sightings of American golden-eyes, bufflehead, surf scoter, ruddy ducks, greater and lesser snow geese, Canada geese, whistling swans, hooded and red-breasted mergansers, black ducks, gadwalls, baldpates, blue-winged teal, shovelers, lesser scaup, canvasbacks, wood ducks—in fact, most of the Atlantic Flyway species.

The Bahama Islands provide breeding habitat for the Bahama pintail, of course, and winter habitat for pintails, shovelers, lesser scaup, bluewings (and greenwings!), ruddy ducks, redheads, and so on. The ubiquitous mallard also winters here, as well as almost everywhere from New England southward. Moreover, just as the Atlantic Flyway is visited by European waterfowl, many ducks and geese of the eastern seaboard are occasionally found on European sojourns.

Still, none of this explains seemingly aimless flights that have been sighted far out on the Atlantic. Some observers have speculated that perhaps overpopulation of a given species in a small belt of habitat may, in unusual instances, produce a lemminglike flight to nowhere. This must occur rarely, however, for overpopulation is not a frequent problem. On the contrary, the problem is the maintenance of healthy numbers in spite of the attrition of habitat as man paves a continent, and in spite of nesting-season droughts, blights, and outbreaks of such diseases as botulism.

*Blue-winged teal migrate earlier and farther than any other American ducks.*

*Green-winged teal winter in Atlantic Flyway states as far north as Virginia.*

Hunting is important in controlling the populations of many water-fowl species, some of which would otherwise become an agricultural scourge. But there are other factors, good and bad. For example, New Jersey's Department of Environmental Protection has estimated that between 1950 and 1970 the state lost an acre of wetlands per hour. A strenuous effort is now being made to halt or even reverse that dismaying loss. Other estimates have put the total fall population of Canada geese at three million or more, with perhaps half that number heading toward the eastern flyways. But in spite of widespread habitat destruction, there are havens that support more geese now than formerly. For a decade and a half, the number of Canadas in the Chesapeake wetlands has increased at an annual average of six percent, primarily because of the Delmarva region's corn farming, plus the protection of the bay and the area's refuges.

It may be that some three million Canada geese embark on the south-ward journey each fall, headed for wintering grounds which, in all their vastness, cannot feed more than half that number. But no more than half that number need be fed, for no more than half will arrive. Some of the others die of natural causes, including age, some die in accidents. The relative safety of nocturnal flight may be shattered by an invisible power line, or by a metal roof glinting like water in the dim starlight when clouds smother the moon's surer illumination.

More are killed by predators. Man, most efficient of predators, too seldom realizes that his atavistic sport, his predatory instinct, is included in nature's balance. Were it not for the insanity of gunners who shiver away the dark in wind-lashed blinds, the geese would swarm like locusts over the southern wetlands in multitudes doomed to slow starvation. Yet a bird may run the continent-long gauntlet of shot for season after season until overtaken not by the reverberating crash of gunfire but by the silent death of the aged.

Each autumnal journey during the life of that bird is an achievement of puzzling magnitude. Even when buffeted by winds, both ducks and geese can cover vast distances at a steady, effortless speed of forty to fifty miles an hour. Canvasback ducks can attain an air speed of more than seventy miles an hour when pursued. The greater part of migratory flight takes place at altitudes below three thousand feet, but snow geese have been observed at twenty-six thousand feet. Such aerial prowess is made possible not only by a remarkable metabolism but by remarkable musculature. For example, a bird's pectoral muscles, contracting to move the wings, simultaneously push against the ribs to pump air into the lungs like an artificial respirator. It has been said that ducks cannot lose their breath because they fly into it.

The navigational ability of wildfowl is another mystery which is being slowly (and perhaps only partially) solved by the banding of trapped birds and the recording of where and when these birds are shot or recaptured. It is a relatively little-known fact that the first American banding experiments, perhaps the first of importance anywhere in the world, were performed on the Atlantic Flyway, near Philadelphia in 1804. The experimenter was none other than that avid gunner of the Atlantic and Mississippi flyways, John James Audubon.

Unfortunately, no naturalist renewed the experiment until 1899, when a Danish schoolmaster named H. C. C. Mortensen began capturing and banding teal, storks, and starlings, evidently without knowing that his idea was antique. Three years later the research was taken up in this country by Dr. Paul Bartsch of the Smithsonian Institution, and in 1909 the American Bird Banding Association was organized. Its program was absorbed in 1920 by the United States Biological Survey, predecessor of the Fish and Wildlife Service, which still carries on the work. By now, literally millions of birds have been banded. The research is intensive and continually rewarding. It was the Service that published *Migration of Birds,* by Frederick C. Lincoln, whose pioneering investigations have been previously mentioned. Dr. Lincoln had charge of the Service's migratory studies, and his findings (based on band returns and since confirmed by additional returns) established the concept of flyways and helped to map them. As a direct result of his contribution, the flyways have been successfully used since 1948 as the basis for the annual review and setting of wildfowling regulations.

As to what has been discovered thus far regarding navigation, it is now known that birds navigate by the sun and stars, correcting course as the journey progresses to take into account the shift in star positions and the sun's path. They also employ landmarks both large and small.

A mighty river may draw a flight forward for many miles, until a mere hillock of shoreline conformation signals the approach to a resting and feeding site. Along the densely settled stretches characteristic of the Atlantic Flyway, wildfowl have actually been observed to follow highways, and they seem to use city skylines and lights as beacons, too.

In 1966, an investigator named F. C. Bellrose reported on this use of landmarks in orientation and navigation, and the findings were confirmed in a 1970 report by Bellrose and Robert D. Crompton for the Illinois Natural History Survey. In emphasizing that "landscape features are important factors," the report relies on direct observation as well as banding: "We have frequently seen migrating ducks change directions upon sight of prominent landscape features."

This does not, of course, reveal how a duck or goose on its first migratory flight recognizes landmarks or translates the map of sun and stars. The miracle is, in part, a matter of instinct, interpreted lately by some animal behaviorists as analogous to Jung's theory of "racial" or "collective" human memory. According to this interpretation, migratory birds possess a vague, unconscious, inborn knowledge of ancestral routes and destinations, a physiological programming for the first long journey. In experiments with birds hatched and reared in captivity, then released to migrate without guidance from elders of their kind, wildfowl have managed to arrive on their ancestral wintering grounds. Still, what this may prove is no more than conjectural, for it is argued that such birds simply search for wetlands, where they inevitably find and join flocks of migrating wild birds. Moreover, studies such as that of Bellrose and Crompton prove conclusively that ducks become much more adept at homing with each repeated migratory experience.

Along the way they guide one another, reinforcing directional memory. And when visibility is poor, it is believed that they may chatter not only as a form of reciprocal encouragement but as a method of echolocation. The echoes bounce back to them from various terrain features and their sensitive hearing translates the messages of this natural radar system into directional instructions. Their voice apparatus seems to function without interfering with the breathing rhythm, so that it is always functional when needed. The man who hears the cries of geese far out over Delaware Bay on a foggy morning may be listening to birds probing an invisible map for a remembered cove.

A sudden early blizzard or a freeze that leaves too little open water in the north country will, of course, send throngs of wildfowl scurrying southward, but since they are likely to dally along the way, the migration schedule becomes a self-correcting process. It is the synchronization of the homing ability with this miraculous sense of timing that accounts for the entrance of ducks and geese upon the right stage in prompt response to nature's cues. Recent studies indicate that migratory birds are so sensitive to barometric pressures and other climatic indicators that they can "feel" changes in weather conditions about twelve hours in advance, and this weather-forecasting instinct helps them to arrive safely at various destination points along the flyway. The sense of timing is so acute among coastal birds that, after flying inland, they return to shore for feeding at just the hour when the tide is right.

A different and even more awesome timing ability governs the seasonal onset of migration so effectively that an early blizzard or freeze is merely a small variable in a centuries-old schedule.

It is possible to observe the birds making advance preparations for migration. Mel Evans, manager at the Santee National Wildlife Refuge in South Carolina, where large numbers of Atlantic Flyway ducks, geese, swans, and wading birds rest or winter, has remarked that he can tell about two weeks ahead of time when the spring migration northward will begin. Using the behavior of the birds, particularly Canada geese, as the key, he can predict almost to the day when the major exodus is to start. The geese feed less and less, and they become so restless that they are almost constantly on the wing. There is a great increase in aerial acrobatics, a tumbling activity that is especially marked among the younger birds. Evans refers to these antics as playing, but adds that the playing is of a sober nature. Canadas are not ordinarily among the most playful of wildfowl. Until migration becomes imminent, in fact, their behavior exhibits an almost haughty dignity and orderly discipline.

Evans and many other observers believe that the sudden indulgence in cavorting is a way of exercising by which the birds condition themselves for the migratory flight. It may also be a natural psychological preparation for the sustained activity to come.

But how do the birds sense the approach of migration time? It is no simple matter of warming or cooling weather. A snowfall in July will not send waves of ducks and geese down the Atlantic Flyway from James Bay or Hudson Bay, nor will an abnormally warm autumn significantly delay their departure. Only when the time is right do the multitudes converge along the tidewater flats of those bays, as they have for countless autumns through aeons past. The skies and waters are then speckled with the thousands of blues and snows and Canadas. One can actually see the impatience take hold and surge through the flocks. Their chatter swells, becomes more raucous, until it is almost a palpable element of the air. Little bunches leave their comrades and make brief flights, still tentative but betraying the yearning to be gone. And then, on the environment's appointed day, heralds of the flight lift skyward, leading still scanty formations as they call urgently to the hesitant. As more birds heed the beckoning cries, the urgency subtly attains a purposeful quality. The tidal waters ooze over the flats as on every other day; nothing appears to have changed. Yet geese soon becloud the skies, formations blending with other formations, chevrons bulging into amorphous skeins.

For a little while the cries resound across the waters before melting away as imperceptibly as the tide itself. It is a seasonal farewell, and when the last echo finally subsides, an ineffable emptiness strikes the flats.

Patient observation and experiment have partially solved the mystery of this exquisite seasonal timing, for it has been proved that migratory birds possess a glandular specialization known as the biological clock. "Biological calendar" would perhaps be a more felicitous term for the seasonal phenomenon, but no matter. It is founded in the sexual drive which assures the perpetuation of species, and is powered by solar energy in the form of light. Among creatures that mate seasonally, there is a genital and hormonal shrinkage to the point of dormancy during the long nonbreeding period while all biological powers are concentrated on survival in a wild environment. Light governs the waxing and waning.

In describing the cycle, it is natural to begin with the spring awakening, when most wild forces come to life. Light stimulates the seasonal development of the birds' sex glands and hormones. As winter ends, there is a daily lengthening of the light over those Santee marshes, awakening the sexual drive, catalyzing the biochemistry of reproduction. As this primeval drive strengthens, the birds grow restless and the restlessness swells until it culminates when the birds strike out for their northern breeding grounds.

Scientists have confirmed this explanation by exposing captive wildfowl to increased hours of artificial light. From year to year, the variation in the hours of daylight during a given period at a given latitude are so slight that the biological clock of each species is superbly accurate. However, the clocks of all species are not set to the same increases and decreases of light. Evolution—nature, if you will—has meticulously staggered the timing of migrations so that overcrowding does not bring disaster along the corridors of flight and rest. This

*Timing of migrations permits long hunting seasons without overharvesting.*

staggering applies to fall as well as spring, spreading the total migration over several months. It is why the gunning season can be satisfyingly long without devastation to the wildfowl populations.

As every hunter on the Atlantic Flyway has found, there is a considerable lapse of time between the first September skitters of teal and the waves of hardy scaup, among the latest of migrants. During the interval there is a steady flow of other species down the flyway. Pintails start early, often at about the time of the teal, but their journey is more protracted, as is that of other ducks that begin to wing southward in September—ruddies, mallards, blacks, redheads. These birds are followed or accompanied in September or October by wood ducks and shovelers, and October also sees the ringnecks, canvasbacks, and American golden-eyes leaving the north. The Barrow's golden-eye is headed for winter quarters in October or November, as are the old-squaws and the American and red-breasted mergansers. Scaup are not unique in their indifference to cold, for such ducks as the harlequin, bufflehead, and American eider remain in the upper latitudes until November and, on rare occasions, even a bit later.

The schedule is both complicated and eased by the fact that all birds of a single species do not migrate during one short period. In the fall, as the equinoctial control of the biological clock is reversed, the sexual and family-raising drives have abated, replaced by other urges. Though the adults of some species still exhibit a somewhat protective attitude toward the young, the immature birds are now strong flyers. The mature birds have regained flight feathers after the moult, and all are capable of extended flight, yet it is to their advantage to continue feeding and resting as long as possible before migration. The diminishing daylight seems to warn them of a coming scarcity of food, at the same time evidently working subtle changes in their biochemistry to prepare them metabolically for the efforts of the journey. But the effects of diminishing light are modified by other factors.

Generally speaking, the birds that subsist on molluscs and fish, and those that frequent the big waters migrate a little later than puddle ducks, and the reason seems obvious: their habitat is not soon frozen over, barring access to food. Similarly, mallards nesting about the eastern Great Lakes have no reason to leave as early as those on the windswept prairies of Manitoba.

Even the black duck, a species encountered in significant numbers only on the Atlantic and Mississippi flyways and having a surprisingly limited breeding range, exhibits a diffused pattern of migratory timing. Since the black is of extreme importance to the gunners of the Atlantic Flyway, it merits special attention here. This duck nests only in the East, though pockets of the range extend into Manitoba. The breeding area sprawls from the upper shores of the Great Lakes northward and eastward through Ontario, Quebec, Labrador, Newfoundland, New Brunswick, and Nova Scotia, with some small coastal concentrations scattered as far south as the Middle Atlantic states.

*Ducks Unlimited map outlines chief breeding area of black ducks.*

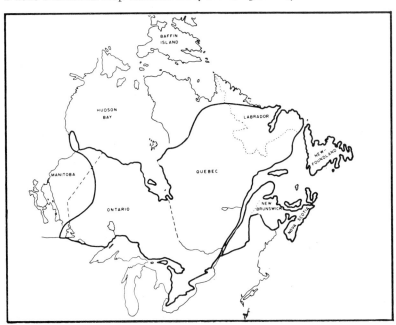

The autumnal migration of those in the lower part of the range tends to begin substantially later than the procession from the North. Some of the southern black ducks, in fact, do not migrate at all. They have no need to—and the absence of migration is probably somewhat more common among wildfowl than is generally supposed. Occasionally a little group of Canada geese or mallards or other birds will gain the ability (for still unknown reasons) to ignore the nomadic urging of the biological clock if they have settled on some congenial slough. This probably occurs with greatest frequency where seasonal extremes of light and weather are not so pronounced as elsewhere—along the middle and lower stretches of the Atlantic Flyway, for example. On a Virginia estuary or a Delaware tidal flat, a little group of Canadas simply settles down to a lazily sedentary existence forevermore. There are also whole varieties of wildfowl that are virtually nonmigratory even in areas of greatly different habitat and climate: the white-cheeked goose of Alaska, a distinctly localized subspecies of the Canada; the mottled duck of Louisiana and Texas, the Florida duck which is common only on the lower tip of the Atlantic Flyway.

*Echolocation may help lead geese to landing site if visibility is poor.*

As to the blacks, they do have a strong migratory urge, whose timing is dictated in some degree by the nesting area. In spite of these variations, Arthur Cleveland Bent managed to amass the average departure and arrival dates for blacks and other major species in his *Life Histories of North American Wild Fowl*. Since the two volumes of this work were originally published as Smithsonian Institution *Bulletins* in

1923 and 1925, when banding on a large scale was still quite new, the accomplishment is outstanding. Bent states that September 30th is the average arrival date for black ducks in the area of Alexandria, Virginia. Those that continue southward first rest and feed for some time, and blacks aren't usually seen around Mount Pleasant, Charleston, the Carolina Bays until October 22nd. Later, some of them leave South Carolina for still warmer climes, and good numbers of black ducks, according to his compilations, are encountered in Wakulla County, Florida, in mid-November.

A bewildering complexity of patterns emerges as Bent then lists some average dates of major flight departures: Montreal, November 6 and 14; Ottawa, November 7 and 21; Prince Edward Island, November 13 and December 8. Many black ducks, then, are leaving the Canadian waters long after others of their kind have arrived in Virginia and the Carolinas. What seems to be a contradictory listing is further evidence of the connection between breeding area and migratory schedule for any single species, compounded by tarrying along the way even on the northern segment of the flyway. The complexity of pattern and behavior is underscored when, in a listing of egg dates, Bent includes for the period of April 20 to May 10 nesting sites as far south as Maryland and Virginia.

There is a tremendous need for a continuing accumulation and study of such data—as well as information concerning fluctuations in habitat, population, migratory routes, pollution, and the erosion of feeding and resting sites by coastal industry, shipping, marsh-draining land developers and the like. The importance of all this is made obvious by a few more statistics regarding the significance of the black duck to the Atlantic Flyway. Over 60 percent of the average total duck harvest in Rhode Island each year consists of blacks; over 50 percent of those bagged in Massachusetts and New Jersey are blacks; just under 50 percent of those shot in Maine and New Hampshire; more than 43 percent of the Delaware harvest; over 35 percent of the Connecticut and Vermont harvests; over 28 percent of the Maryland take; 26 percent in Virginia; just under 23 percent in New York; and more than 17 percent in Pennsylvania. Even as far down the flyway as North Carolina—the major wintering grounds for a great many other species—black ducks account for more than 12 percent of the annual crop.

As the case is stated in a Ducks Unlimited brochure, "The preservation of the black duck is critical to the Atlantic Flyway harvest." Ducks Unlimited (an organization whose admirable role in conservation will be mentioned more than once in subsequent pages) must be credited

Pintail Duck

with truly enormous achievements in the enlargement and maintenance of wetland habitat for the black ducks of the Atlantic Flyway (in fact, for all the major species on all four flyways). Funded solely by America's sportsmen, DU has contributed millions of dollars to securing or rehabilitating and preserving marshlands. A typical project is the Missaquash Marsh on the Nova Scotia–New Brunswick border, a drained wasteland until the organization built a dike and water-control system. There are now twenty-six miles of shoreline on this six-thousand-acre marsh, an important producer of blacks and other birds. The project was principally financed by DU's Delaware members, and another was made possible by the organization's Rhode Island members. This is on the Maccan River, ten miles south of Amherst, Nova Scotia, where a two-thousand foot levee has been built. The result is a small but important marshland, two hundred and ten acres with a long shoreline that contributes to the flights of black ducks heading down the flyway. In the spring of 1971, Ducks Unlimited opened a regional office in the Maritimes to coordinate new operations, initially costing well over $200,000, in the heart of the black-duck country. As this is being written, eight new projects are being maintained, developed, or planned in Nova Scotia, Prince Edward Island, New Brunswick, and Quebec. In this region alone, the organization maintains over ten thousand acres of wetlands which help to assure the future waterfowl population of the Atlantic Flyway.

Another species peculiar to this flyway—almost unique to it—is the American brant. This diminutive marine goose winters chiefly from New Jersey to North Carolina, with the greatest concentration in Virginia and North Carolina. In connection with the autumnal arrival of brant, one thinks immediately of such traditional gunning bays as Brigantine and Barnegat, or of the shores of the Delmarva Peninsula. Yet brant, like other migrants, can be capriciously nomadic. Although they don't often stray very far off course, Bent recorded (probably with some effort to restrain the delight of serendipity) that sightings had been recorded in Michigan, Nebraska, and Wisconsin. Once, in

November of 1876, a brant was observed in Barbados. And the birds are occasionally seen on the Pacific Coast where a relative, the black brant, is alleged to be the exclusive representative of the clan.

Most northerly of all our geese, the American brant nests along the coast of Greenland and has more extensive breeding grounds on Ellesmere Island, well within the Arctic Circle. The short polar summer forces an early migration which descends into the Gulf of St. Lawrence before September ends. From that point on, however, progress is so leisurely that brant aficionados of the Atlantic Coast are assured of seeing the flights continue for a couple of months. Long, teetering lines of brant weave and undulate over the horizon, their quick wingbeats sometimes giving the impression of desperate escape from a sinking sky, a marked and touching contrast to the majestically effortless progress of the Canada geese as the brant sputter along toward wintering grounds that will not often be reached before mid-October or early November.

Apart from oddities of flight, the brant has other characteristics that are somehow touching. As with Canadas and other geese, the male stands guard close by while the female incubates the eggs. But a guardian brant lacks the imperious presence displayed by a guardian Canada. On the stark arctic tundra, it seems miraculous that a ground nest of moss and grasses, though lined with down of maternal breast, can sufficiently protect those three or four or five fragile, creamy white eggshells that hold the embryos of another generation.

Brant do not graze upon the dry land as do Canadas, snows, and many other geese but they, too, are essentially vegetarians and, unlike so many of the marine species, an epicurean delight. Their chief food is eelgrass (*Zostera marina*), which thrives in shallow bays and estuaries. This single staple is even more vital to the American brant than it is to the black brant of the West Coast. A hunting man on the Atlantic Flyway dwells upon a fact like that, for it demonstrates the fragility of a coastal ecosystem which, to the uninformed, appears too rugged to be gutted by any calamity short of technocracy's inclination to spread oil upon the waters.

Because brant have long been recognized as the most succulent of waterfowl, the batteries of the market gunners were trained on them; they survived even that. And then, in the early 1930s, a fungus struck the eelgrass.

The blight was severe along the entire Atlantic Coast and as far away as France. Starvation ensued. Before the famine, there had been more than a quarter of a million brant on the flyway. By 1935, there were

*41*

*Since eelgrass has returned to coastal bays, brant are again thriving.*

perhaps twenty thousand left—about two percent of the former population. Birds more flexible in their migratory habits might have sought new habitat, as almost all species of wildfowl do in the event of catastrophes. But then, birds more flexible in their migratory habits would have to be more flexible in their dietary requirements—in which case there would be no need to seek out a new migration route.

What happened was a classic example of survival of the fittest, which in this case meant survival of the most adaptable. The few remaining brant, hardiest of their lineage, eked out an existence on substitute forage—sea lettuce and similar vegetation.

The eelgrass very slowly revived, and with it the American brant population. As recently as the summer of 1961, Van Campen Heilner, in an article appearing in *The American Gun,* said with some justification that "today the word 'brant' is almost forgotten in the lexicon of waterfowl gunners." The article was not, however, an epitaph; on the contrary, it was a paean to the miraculous return of brant to the Atlantic Flyway. Heilner continued, with regard to *Branta bernicla hrota:*

it is the name for a goose of diminutive size and succulent flavor that swept down the Atlantic coastal flyways in hordes each autumn to gladden the heart of every wildfowler with stamina enough to face the stinging winds and pelting rain of a wild northeaster. Good brant weather has always been bad weather, with curling crests of raging water smashing over sand bars, salt spray and stinging sleet driving into the blind. No fair weather bird is the brant, for it continues to fly or lies happily out to sea until the storms bring it into gun range. Then, flying low, it seeks shelter in the coves. . . . The greatest concentration of brant that I have ever seen was off Hatteras Inlet, North Carolina, in the days before the disappearance of the eelgrass. It was early November. . . . The bend of the beach near the inlet was black with them. . . . A steady stream of brant poured out of that bend for more than an hour.

Today, descendants of those birds still ply the Atlantic Coast, helplessly loyal to their ancestral foraging and migrating instincts. They are multiplying again, their numbers growing, their survival and renewed prosperity another beautiful mystery of this migratory corridor.

# Chapter Five

# SECRET LIFE OF THE SALT MARSH BARRENS

After several years of scrimping and cutting vacations short, a hunter from New York had finally managed to arrange his fall business schedule so that he could grant himself several week-long tours north and south on the coast he loved, exploring uplands and lowlands that he had never known well enough and probably never would know well enough to satisfy his curious yearning.

The season had begun inauspiciously in September, when he had accepted a weekend invitation from a Connecticut friend who wanted to introduce him to rail shooting on the marshlands sprinkled about the Connecticut River and Nells Island, near the mouth of the Housatonic. During the drive up, he had let his mind dwell on the somber and paradoxical combination of serenity and intensity in the Thomas Eakins paintings of sportsmen and their guides "pushing for rail." He was under no illusion that the habitat or the hunt would be quite like the Eakins milieu of 1874, but he hoped that the same palpably quiet atmosphere might prevail, and in this he was not disappointed. Neither was he disappointed in the suspense of waiting for the reluctant flush and short, erratic flight of the shy marsh birds, their legs dangling as they skittered over the tall grasses. Other elements of the hunt were not, however, quite so perfect.

*Ruddy turnstone joins semi-palmated sandpipers foraging at tideline.*

*Painting by Thomas Eakins exudes atmosphere of rail hunting in 1870 s.*

His own preparations had not been as meticulous as he might have wished. Though he was armed with a cylinder-bored 20-gauge gun and had managed to acquire some Number 10 and Number 11 shells, classic choices, business had prevented him from timing his hunt to take advantage of the highest flood tides on a day of the full moon or new moon. The guide had been reasonably competent, but with the water a bit low, poling was difficult. The art of the pusher has always been a demanding one. He must know the productive pockets of marshland, and he must pole the little shallow-draft boat along at just the right speed to prod unwilling birds into flight within range. Progress must be smooth and steady, for the gunner has to stand, ready to shoot fast over the tall reeds. Balance is essential. It is likely that truly great pushers have been in short supply ever since the sport reached its zenith of popularity in the last quarter of the nineteenth century.

Still, the New York hunter had managed to shoot a few of the whistling, whinnying sora rails, and had seen a couple of the larger clapper rails as well. Besides, he anticipated another try at clappers later on, near the mouth of the Cape Fear River in North Carolina, where the season stretches from the beginning of September through the first week in November. There, on the tidal marshes near Southport, one can find the king rail and the Virginia rail as well as the sora and the clapper, known locally as marsh hens. So he was not unhappy about his season's unimpressive beginning; it only whetted his appetite for longer coastal sojourns.

Early in October he shot woodcock in New Brunswick and Maine, and from a cramped sneakboat on Merrymeeting Bay he bagged a few teal. Then he wandered a bit on the chain of tidal marshes between Kittery and Portland before driving south to Portsmouth, below the New Hampshire line.

*King rail pecks amid lush growth at edge of southern salt marsh.*

He walked the bleak barrier beaches which shield the marshes from the roaring surf, and even on the rustling sands where provender seemed scarce to his anthropocentric eye, the birds were plentiful—not just gulls, petrels, shearwaters, the expected ocean denizens, but grebes and shorebirds: plover, yellowlegs, sandpiper.

The Eskimo curlew and the Labrador duck were no more to be seen, of course, having succumbed long ago to the rapacity of an unlamented bygone era, but there were American coots and "sea coots"—scoter that he had gunned from sculls and gray double-ender lobster skiffs in the Maine coves and once from an unwieldy dinghy on Long Island Sound. He recalled how, during the early part of a past season, they had been so innocent as to decoy to lobster buoys and net corks. They would even toll in over an anchor-line float, though they did seem to acquire caution after a few volleys.

If succulent vegetation appears scant along the pebble beaches, breaker-smashed rubble, and cliffs, there is no mystery about the source of food for scoter. In one study, the stomach contents of 819 white-winged scoter yielded an analysis of 94 percent animal matter. Molluscs alone—rock clams, oysters, blue mussels, scallops—comprised 75 percent. Some crustaceans, with a few insects and fish, also provided a significant proportion of sustenance, while eelgrass, bur-reed, and miscellaneous plant foods made up only 6 percent. Yet Van Campen Heil-

*Bay ducks and shorebirds furnished themes for many Currier & Ives prints.*

SHOOTING ON THE BAY SHORE.

SHOOTING ON THE BEACH.

*Coots (left) cruise about on salt marsh; eiders favor off-shore reefs.*

ner's excellent work, *A Book on Duck Shooting*, concludes a passage about hunting scoter with this declaration: "And when you take me home that night I want a big red-hot, steaming plateful of good old coot stew!"

There is no denying that even those waterfowl (or most of them) which rank low on epicurean lists because of a large intake of fish and shellfish can yield delightful table fare. Most connoisseurs utilize only the breasts, carefully marinating them; others combine the use of a marinade with a "throw-away" stuffing intended to draw off any trace of fishy flavor. By a number of means, man has elevated even the lowly among waterfowl to a significant rank in terms of his carnivorous appetites as well as his sport.

The surf scoter imbibes a bit less animal food than the white-winged variety, and a trifle more vegetation, including pondweeds—a reminder that even "marine" waterfowl are to some degree sustained by the sweet-water wealth of the inland marshes as well as the brackish and salt-water staples. The American scoter eats about as much animal food as the surf scoter, supplementing its diet with pondweeds, muskgrass, and miscellaneous plant food.

Curiosity had led the New York hunter to delve into the reports of game biologists, and he knew enough about the foraging practices of coastal birds to understand that the stark shores, dark waters, and especially the flat, monochromatic ochre of lifeless looking salt marshes appear barren only to the unpracticed eye. They are delicate but rich ecosystems teeming with hidden resources, with life-giving life. Nevertheless, he felt a mild astonishment (or reverence, perhaps) at the ability of such birds as Canada geese, scaup, golden-eyes, and buffleheads to flourish on the coastal marshes and open waters as he prowled the estuary of New Hampshire's Great Bay.

46

Riffling through the pages of Kortright, he pondered on some revealing analyses of foods eaten by these birds. In an examination of 395 American golden-eyes, scientists had found that 74 percent of the diet consisted of animal matter, again consisting of molluscs, crustaceans, fish, and insects, supplemented by pondweeds, wild celery, spatterdocks, grains, and bulrushes. The percentages were not greatly different for 81 Barrow's golden-eyes. A sampling of 282 buffleheads revealed a slightly higher percentage of animal foods, particularly insects, similarly supplemented by various aquatic and terrestrial plants. Most of the diving ducks are either predominantly vegetable eaters or animal feeders, but the greater scaup is a borderline case whose diet consists of about 50 percent vegetation and 50 percent molluscs, insects, crustaceans, and so on. Among 752 birds examined, the vegetation consumed included significant percentages of pondweeds, muskgrass, water milfoils, sedges, wild rice and other grasses, and wild celery. The diet of the lesser scaup includes more vegetation—about 60 percent. Over 1,000 of these birds were examined, and they had eaten significant percentages of pondweeds, wild rice and other grasses, sedges, bulrushes, wild celery, muskgrass, water lilies, coontail, water milfoils, smartweeds, and arrowheads. Samplings of this kind usually reveal puzzling exceptions, ducks which have been feeding exclusively on animal matter or exclusively on vegetation. However, such anomalies do not indicate a preference but merely the availability of one food or another in the most recently occupied habitat.

Some of the plants thrive best in or around fresh water, but others flourish in brackish or salt marshes. Such wetlands may often appear sterile except for reeds, high grasses, and occasional sedges, for they do not support luxuriant blankets of grains and shrubs. Their waters, too, appear relatively barren, the surface unadorned with lilies and duckweed, arrowheads and pondweeds. Nonetheless, those waters may harbor eelgrass, widgeongrass, wild celery and such (as well as sea lettuce and other aquatic weeds for the species of less fastidious dining habits).

Even the Canada goose, whose love of grain has ensconced the species in the gourmet's peerage, just as the bird's power and eerie intelligence have mythologized it among outdoorsmen—even this great wild goose can pluck sustenance from what seems barren.

Inland the living is easy as the flocks graze amid the stubble of prairie grainfields, glean the fallen grain in autumn, pick the lush green herbage of pasturelands. During spring migration, they may sometimes wreak considerable damage to young wheat, barley, corn, and oats. There is no doubt that on the easternmost flyway's great mid-Atlantic corn belt, with its mechanical pickers spewing kernels and chaff and half-denuded cobs as if the lumbering machines were giant robot poultrymen gone berserk in a sea of hungry birds, the migration pattern has been affected. Greater and greater multitudes of Canada geese, together with mallards and, to a lesser degree, other grain-loving wildfowl, have shortened their southward procession to pass the winter close by the corn stubble of Virginia, Maryland, and Delaware. It is probable that a larger wintering population of birds can be supported in these areas than could have survived prior to the advent of modern high-yield farming practices and the time-saving but grain-sacrificing mechanical harvesters.

There have been complaints from some wildfowlers to the south that their abundance of fowl has been preempted by the mid-flyway corn growers. Cotton and tobacco fields have little attraction for ducks and geese. Corn is the lodestone.

Still, it is unlikely that the more southerly wintering populations have been adversely affected to a perceptible degree. Each wintering haven absorbs the number of birds its habitat can support, just as it has for aeons past; the rest continue southward except for those harvested along the way. And corn is not the sole grain that magnetizes flocks; they can settle like a cloud amid the volunteer wheat sprouting in other stubble fields, or swarm like locusts where rice fields replace the miles of corn.

During midday, Canadas prefer to rest on open water, sandbars, or mudflats—fresh, brackish, or salt. In the marshes there is no need to fast. Long before man's agriculture mowed the forests, filled the swamps, and covered the land with endless regiments of grain, the geese in the marshes fed well on wild rice, sedges, aquatic plants, insects, larvae, crustaceans, small molluscs. They still do. Though they are the most terrestrial of waterfowl and prefer to blanket the cornfields, they also forage as do the surface-feeding ducks, cruising about the shallows, tipping up their tails and feet as their long necks reach down while they pluck the riches from the bottom. The competition of other species is a small matter, for they do not seek the company of lesser wildfowl. And although a few intruders such as baldpates (aptly called thieving ducks or poachers in some locales) may hover about the fringes of a flock of Canadas to peck nervously at leavings, the great geese do not tolerate the intimate company of competitors.

Knowing of these things, the hunter could begin to understand how the birds waxed fat even during the arduous fall migration as they lingered upon the drab marshes and cold, inhospitable waters of New

*In sudden flurry, pintails rise on tidal marsh.*

Hampshire's coast. He knew, too, that unlike an invasion of industry or concentrated settlement by man, an influx of wildfowl is more likely to replenish than impoverish a delicate ecosystem. The birds take only what the marsh can afford to give as nature balances itself. If foods are in short supply, the birds will move on, but before that happens a thickening, spreading crust of ice usually hampers feeding and prods the birds southward on their autumnal journey. Meanwhile, they have enriched the habitat with fertile droppings. Even shed feathers decompose and become constituents of a nourishing ecosystem supporting a complex chain of plant and animal life. Somewhere on an isolated marsh, a bird dies of age, disease, accident, or predation. Its flesh and bones enhance the nutrients of the wetlands; nothing is wasted, not even the dried bits of eggshells flicked about by the winds in the north.

One day the hunter tired of spreading huge rigs of decoys on the open New Hampshire waters—three or four times as many as needed on a small fresh-water marsh. While he did not much believe in the old "fish hook" or "crescent moon" spreads, he had expended much energy placing his decoys in a more or less pear-shaped arrangement—with his boat, camouflaged by a matting of bamboo and reeds, anchored at the pear's stem, on a known flight path. He carefully left an opening amid the decoys in front of the boat, a space where the birds could pitch in. And he situated the whole arrangement so that ducks would fly over the mass of stool, heading upwind toward the opening and the camouflaged boat, as they set their wings and attempted to alight.

Without a guide to perform the manual labor, this kind of waterfowling involves a great amount of heavy work even for two gunners. Small wonder that he tired of it after a week and, with a friend, went "cricking." The excitement of trying to anticipate the unexpected is a special joy of "cricking"—jump-shooting along the tidal creeks meandering through the marshes. Like almost every American wildfowling technique, it originated among the early settlers on the Atlantic Flyway and has since spread to all regions.

On this day, the hunter and his friend were not using a boat. It was low tide, the safest period for such a venture on a coastal marsh, and even chest-high waders seemed an unnecessary encumbrance, so the men chose to plod along in hip boots, crouching sometimes and creeping over hummocks to peer at potholes and creek openings where perhaps a few black ducks might be caught unawares.

The hunter from New York was thinking about various, perhaps trivial, bits of learning he had acquired along the flyway: how previously sterile manmade ponds had yielded sunfish and perch after they had been visited by migrating ducks for several seasons—ducks that must have carried fertile fish eggs on their feet, strengthening the ecological links of their migratory chain—and how a northeast storm brings the birds in close to a floating blind on the New Hampshire coast but a northwest wind is prayed for by the wildfowlers down on the Eastern Shore, along the Maryland wetlands where he would soon be hunting.

His mind was on such matters as he sloshed through ankle- and shin-deep marsh water, grasses hissing against his boots, while he looked for some perceptible demarcation in the reeds and blades, some line to reveal a tidal creek he knew to be close by. He wondered if perhaps it would have been wiser to don waders for this excursion.

Abruptly, his left boot plunged downward and found no support. Quickly but much too late he recalled a line in some outdoor magazine to the effect that "in unfamiliar territory you are likely to float your hat by stepping into a tide-hidden creek."

Afterward, as he watched the musky steam rise from his clothes before his friend's fireplace, he decided it was time to visit the warmer Eastern Shore.

A day in early November, colder than he had anticipated, found him in Maryland, just below the Delaware border, with another friend. They hunted quail that afternoon, ambling with a setter and a Brit down a large soybean field bordered by woods and a fallow pasture overgrown with high brush and scrubby pine saplings. On today's manicured farmlands the bobwhite quail is at best an impoverished tenant, scratching hard for a living and huddling in scant cover. But in areas where the farmers leave some of the land fallow, permit high, rough edges to grow, and keep some brushy woodlots in reserve, the quail covey still dominates the mid-Atlantic and Southern uplands.

The farm in this instance was admirably ramshackle. The very rows of soybeans wavered like sloping, broken lines of black ducks, and the gray-brown pods hanging heavily from the angular stalks were interspersed with jutting clumps of weeds. The setter bounded along an edge, brushed by tendrils of Virginia creeper, then checked his gait and walked a few stiff-legged paces, froze, dropped almost to his belly and gazed intently into a brushy corner. Before the Brittany spaniel could brake her run to honor the setter's point, a dozen quail erupted, feathered balls of shrapnel flung into the woods.

The Marylander brought down one bird and the setter fetched it to him, a plump, mottled cock with a very white chin patch. The New Yorker, not yet accustomed to such wild flushes, had failed to get off a shot. The next hour was spent in a quest for singles, scrambling over blowdowns, through little green hells of cat's-claw and holly and devil's-club.

The yap and yelp of beagles could be heard to the east, where several farmers were rabbiting. On previous visits to this area, the hunter from New York had heard the bugling cries of foxhounds, and now he was only mildly startled when a fox stepped daintily from a tangle directly before him. For an instant the animal stood facing him, confused perhaps, momentarily stripped of the vital instinct to flee. Had it been a raccoon prowling in broad daylight, the hunter would have fired by now, but with a fox there was more to think about than a predator's villainous role as destroyer of eggs, fledglings, and even adult birds.

*Close to southern wetlands are upland fields where bobwhite reigns.*

The men of this region liked to pursue the fox with hounds, but almost invariably after being run to ground the fox was permitted to live so that it might lead the chase again on some brisk morning. Many of the local farmers would be delighted to see a bird hunter kill a fox on their land but others would not take kindly to the final demise of their phoenixlike quarry. The owner of this acreage might well be a devotee of hound music; the New Yorker had met him but once and did not know.

As the fox wheeled, its dense coat rippled luxuriantly. This was no red fox, descended from English stock imported in the eighteenth century for the entertainment of the colonial gentry, but a gray fox, a native. For some ungrasped reason, not thought out, the recognition stretched the hunter's hesitation, as if the predator—together with the upland birds and mice and waterfowl on which it preyed—had more right to patrol this ancestral domain than man the newcomer had to intrude upon it. The animal streaked away, low to the ground, its brush straight out.

"Hey!" the hunter called to his companion. "There's a fox!"

"Shoot it!"

But of course the fox was gone.

Beyond a waist-high jungle of creepers, the decaying carpet of leaves took on a deeper brown hue where the earth sloped toward an intersection of two long-forgotten irrigation ditches. Here the hunter from New

*Gray fox is native to Atlantic Flyway's woods and fields.*

York slipped his third quail into his game pocket and was about to trudge on past a copse of swamp maples into a scraggly stand of alders when he heard a small, tentative, nasal chirp somewhere ahead. "Peent." Instinctively he raised his gun, seeing nothing but a brown clutter of foliage and sapling poles. He heard the woodcock rise before he spotted it, the whistle of air twittering through the three outer primaries of each wing as the bird dodged upward out of the twisted screen of branches.

Even as he pulled the trigger, the hunter knew that he had swung too late and was far behind the bird. As he walked on, he mentally toyed with his only alibi—that both dogs had been off to the left and had failed to scent the woodcock, so he had not been forewarned. As paltry an excuse, he reminded himself, as being caught off balance. He resolved to be more alert as he trudged on, and there were plenty of warnings now: the whitewash spatters dropped by woodcock, the borings where the birds were drilling for earthworms in the soft bottomlands.

"Nice to know they've arrived," his Maryland companion said. "I hunted here just yesterday and there wasn't a sign of woodcock. Might be a few natives around but these must be flight birds. Strange, isn't it, how they just show up one day? I mean, it seems like a tough way to migrate, even if they do take their time getting down here. Flying at night like that, you wonder how they find the good pockets to come into. You hit the woods one morning and they're all settled in."

"I don't know," the other hunter said, "I guess it's no stranger than how they manage to make all that distance on those stubby wings. It's like bumblebees. You'd think they'd never be able to get off the ground. Well, let's cover this woodlot."

Though woodcock give off a lighter scent than quail, the dogs were onto them now and pointing them well. Several bounced from the edges of an old orchard. When one rose, the hunters watched for a second bird, for they often seemed to feed or hide in pairs, and once the Marylander scored a double. Just as the whiter, brighter face of the male bobwhite differentiates it from the female, so does the longer bill of the female woodcock set it apart from the male, and the men found that they were shooting about equal numbers of cocks and hens. At dusk they had not quite bagged their limits but they were satisfied.

That evening, after the birds were cleaned and the drinks poured, talk flitted from woodcock to migration, alighting amid rafts of anticipated ducks, and it was decided to rise long before dawn. Only a few miles from the tumble-down farm with its quail and woodcock, the Trans-

*Woodcock—unique "shorebird of the uplands"—is among flyway's important migratory game species.*

quacking River ("Yes," the Marylander said, "that's what it's called.") winds out into Fishing Bay.

On the side of the road was a small, mud-blanket parking area where the state had built a wood and concrete launching ramp. An electric torch danced in the dark as the New Yorker tried to play it on the trailer hitch and winch, wondering if he was furnishing any help at all. It took only a couple of minutes, however, for the Marylander to slide his flat-bottomed sixteen-footer into the water. The two men clattered aboard, stumbling over gear, knocking decoys out of their bushel baskets. On the third pull of the starter cord, the big forty-five horse-power outboard sparked to life and the Marylander climbed forward to kneel at the steering wheel.

At first the shoreline was no more than a low black-on-black silhouette and he seemed to feel his way rather than to see the familiar channel. The boat slapped and rolled against a steady surface chop where the river mouth opened onto Fishing Bay. Several miles down the windward side, on a ribbon of state-owned marshland open to the public for hunting, was a point where a slat-and-rush blind had been constructed. On the water in front of this the hunters trailed the decoys, some in tandem, out on their anchor lines, carefully avoiding the motor's propeller with the streaming cords. Nearest the blind and slightly to the right of it—aloof from the other decoys—a dozen plastic Canada geese rode the swells. Farther out and to the left were perhaps a dozen and a half plastic canvasbacks and scaup, and a few black-duck blocks bobbed on calmer water nearer to a crescent undercut of bank.

False dawn had snuffed itself out unnoticed during the labor of floating the blocks, and the sky was paling by the time the men had hidden the boat and settled themselves in the blind. The New Yorker glimpsed a scurrying dark form, just disappearing among the bases of the tall, thick grasses to the rear. A muskrat, he supposed.

Black coffee. A cigarette. Cold wind.

The sky cleared.

"Wind's in our faces, and the only other points around here are private," the Marylander said. "The birds'll have a hell of a time trying to come into this rig. Well, those are the breaks, I guess."

They could feel the tide-driven pulse of the waves, drumming steadily right under their feet, undercutting the bank, insatiably eating away this little point which would eventually crumble, retreat, reshape itself. It was cold and the time went slowly as the men peered at a cloudless, birdless sky. The New Yorker wondered aloud if the tide and wind didn't constantly erode the marshy shoreline.

It did, of course, his companion said. Some years ago there was farm-land almost out to here, where they were sitting. But the bay's salt water ate away at it, in a slow but unyielding process that had been altering the shore for centuries, and seeped back into the land, raising the water table here, turning once dry and solid earth into a vast tidal marsh, creating small, brackish potholes where the prairielike crust caved in. The undermining action of burrowing muskrats had probably enlarged some of those potholes, but weathering and the subterranean seep alone might replenish such wetlands almost eternally, cutting away, cutting away, as the rivers and tidal creeks brought fresh silt to maintain the delta's balanced fulcrum. Pile and crumble, build and cut, perpetual renewal, unchanging flux.

Still, thought the New Yorker, it appears so barren if the eye takes in only the wide waters before the blind and then, to the rear, what seems like an endless sea of cordgrass, saltgrass, reeds, and needlerush—just about worthless as fodder for the wildfowl. And, indeed, it is quite true that a salt or brackish marsh produces so little food per acre that its vastness is usually what counts.

But he knew there was more to it than that, much more. Just recently he had received a government booklet on the wild duck foods that can be propagated by a landowner who has a marsh or pond on his property. He recognized the salt marsh bulrush mixed in with the almost worth-less growth, rush three or four feet high, with conelike heads bearing a heavy crop of brown seeds—salt marsh bulrush, self-renewing sustenance for ducks, a most important food. And where the fresh sloughs and streams were deep and still enough there was the wild celery. And where the saline current had some force there was the eelgrass rippling toward the surface.

In the fresh or brackish potholes, he knew, there was also likely to be a healthy growth of another important food plant, sago potamogeton. How odd that one thinks sooner of duckweed, which does not attract ducks and, in fact, may shade out submerged food plants.

There would also be dwarf spikerush in fresh, brackish, or even salty potholes. A remarkable example of self-perpetuation, it is sometimes pulled up in quantities by the ducks. They eat the roots, letting the tops float away, and rafted masses of these waste tops often float along the downwind edge of a pond. Yet the plant proliferates and the crop does not seem to have dwindled the following season.

Dwarf spikerush is only a moderately good duck food but it does not compete with the more important bulrush and widgeongrass. The waving widgeongrass grows from the bottom to the surface of sufficiently saline water, furnishing preferred food not for widgeon alone but for redheads, scaup, blacks, gadwalls, green-winged teal, shovelers, oldsquaws, ruddy ducks, and sometimes other species. The voracious birds devour the leaves, stems, seeds, and roots. But this plant, too, has a seemingly miraculous ability to replenish itself. When it appears to have been eaten out, the ducks have left enough so that the grass will reappear.

The real threat to widgeongrass is not the ravening of ducks but the smothering effect of filamentous algae. Fortunately, some fish—particularly mullet—thrive on this algae, thus controlling it. Mullet are frequently stocked in ponds for just this reason. The hunter wondered if their presence in some unstocked ponds might be explained in the same manner as the arrival of perch and sunfish eggs on the feet of waterfowl. Perhaps, but it is likely that most mullet fingerlings are carried by the tides and creeks. However they may arrive, they promptly add their service to the intricate, delicate ecological balance.

As the New Yorker and the Marylander huddled patiently in the cold, damp blind, there was little activity except for a skimming pair of redwinged blackbirds that plunged into the grasses to the right. Later, a Louisiana heron hovered momentarily above the Canada decoys, then flapped its wings ponderously and straightened its dangling legs like a kite tail as it sailed away. Several loons flitted over the water, to the surprise of the New Yorker, who always regarded loons as birds of the

*Canadas winter on marshes—both salt and sweet—along eastern corn belt, where sheltered waters are near grain fields.*

North. And an American merganser passed over the blind time and again, tauntingly. The men refrained from shooting it. A handsome bird, and legal to bag, it is a fish-eater not much sought for table fare. Its worst enemy may well be the fly fisherman, whose proprietary attitude toward trout has been fatal to many a gluttonish merganser.

"Broadbills," the Marylander whispered and hunched down in the blind, cupping his hand over his mouth and purring to them without need of a mechanical call. But the wind remained dead wrong. The scaup saw the decoys, circled in, traded to and fro beyond the rig, and could not pitch in close. Once the New Yorker imagined he could make out the jaunty broad blue bill of a duck in the vanguard of the tightly bunched flock, but he realized that wishful thinking would not bring the birds within range. They finally lit on the water, just beyond the outermost decoys, and there they remained, a bobbing invitation to additional small flocks which soon joined them.

At noon the New Yorker left the blind to ramble back across the marsh, hoping to jump-shoot any birds resting at the lee edges of the little potholes. He found none, but as he crossed a wide, prairielike expanse he saw three black ducks speeding straight toward him. He was in the open, and could only crouch down and hope. They saw him, of course, and flared just as they nudged the edge of shotgun range.

On his way back to the boat he stepped into a muskrat hole, somehow lost his hat, and thought of the accidental dip he had taken in a New Hampshire tidal creek.

Picking up the decoys was a difficult, frustrating chore in the surface chop, but the hunters accepted the cold bite of misery that somehow adds to the joy of wildfowling.

"Skunked!" the Marylander exclaimed. "The little old ladies in tennis shoes would never believe a pair of hunters could be out here without massacring a pile of birds. They're supposed to hang over our heads and wait to get shot, you know."

"I guess nobody told them," the New Yorker said.

"Well, in fifteen years on the Eastern Shore I've only been skunked twice. If you'd been along the other time I'd figure you for a jinx."

That afternoon the Marylander put in a call to a Chincoteague friend, a decoy maker with the implausible name of Cigar Daisey. The incongruity of juxtaposition made the New Yorker wonder if the Marylander had invented that weed-and-blossom name as a joke, but he had read of early English settlers of the Chesapeake Bay area who had been encumbered with such names as Thomas Birdwhistle, John Halfheade, James Tendergrass, and James Wildgoose. Later, upon passing an old churchyard where the largest stone was graven with the family surname Daisey, he was glad he had kept his counsel.

He had also read that the islanders of Chincoteague Bay, steeped in the legacy of outlawed market gunning, could be as secretive and clannish as the descendants of New England's smugglers. The Marylander was telephoning Daisey to inquire about the best gunning prospects for the next day, and the New Yorker wondered if an islander would divulge such treasured information to an outsider. But the Marylander was a friend of Daisey's, and an islander—even an islander far less friendly than Cigar Daisey, who proved to be a most cordial gentleman—likes to accommodate a friend.

The drive to Virginia was short, nonresident licenses were readily available, and the next morning the Marylander and the New Yorker moored their boat at a small island—a "tump," in the Chincoteague vernacular—out on the wide and shallow bay. "It's a good point and public," Cigar had said, and so it was. "The brant're flyin'," Cigar had said, and so they were, as were buffleheads and oldsquaws and occasional cans and redheads and scaup.

Fifteen or sixteen Canada decoys were set out, with a trio of baldpate blocks at the fringe of the rig, and to the left a dozen black-duck decoys. "Should've brought brant decoys," the Marylander said, but the New Yorker later wondered if any more brant could have possibly tolled in over decoys of their own kind. The island was fringed with clumps of myrtle bushes, high beach myrtle that would have offered almost adequate concealment even without the roll of cane fencing which the men unfurled and staked in an opening among the bushes to serve as an admirable portable blind.

In the distance a flock of snow geese passed, white dots across the sky, and then a "V" of five great white swans like aerial sails. Two godwits and later a dunlin flitted over the island. For a while, a Wilson's snipe posed on the bank, a tiny and improbably misplaced lawn figurine. A little helldiver plummeted into a wave and the hunters did not see the bird again.

Though an inquisitive oldsquaw was disdained, there were birds in plenty to shoot at: scaup and redheads that would grace the table, two insistent buffleheads that were missed repeatedly amid squalls of self-derisive laughter—and brant.

Perhaps they are not so majestic as Canadas, nor even so majestic as less imposing geese. Perhaps there is a touch of diffidence in their uncertain strings barely worthy of the title "formation," perhaps their flight looks almost uncontrolled when viewed at a distance. Yet when brant drive in low and fast over the decoys, directly toward the gunner, their curving wingspread commands respect and their purposeful demeanor seems to declare a wild, unconquerable will to survive. The black webbed feet of the leading bird abruptly reach downward, the flock dips, shots crash into the wind. Then there is silence. A man wades into the water, his rubber-cased legs plowing forward slowly, awkwardly. He lifts a limp bird from the surface.

It is vaguely like a miniature Canada though it lacks the white bib and has a short neck and small bill. The bill, head, neck, chest, and forward portion of the back are deep black, relieved only by a broken white crescent of slashings on each side of the upper neck. The back, scapulars, and rump are dusky brown, the sides of the rump white, the underparts and sides ashy gray and white, the tail black above and white below. It is a goose only slightly larger than a mallard, and it is a creature of beauty.

The Marylander examined the area about the rectum of the first fallen brant. If it had been tinged with green, the men would have shot no more brant that day, for a greenish tinge means the brant have been feeding on sea lettuce and other "trash food" which, though it saved them from extinction when the eelgrass was blighted, makes their flesh almost unfit for consumption. It took many years for the eelgrass and the brant to come back after the blight, but there was a turning point in the mid-'50s. After that, the population expanded quickly, and more and more of the birds returned to their ancestral diet of eelgrass. Occasionally, the flights that arrive on a given day have been feeding on the old poverty plants and are marked by the green tinge. But if the tinge is brownish or absent, the primary food has been eelgrass and the flesh will be as succulent as legend claims.

"No green," the Marylander said, and smiled. "They're fine."

Heading landward at sunset, into an oncoming gale, the boat was laden heavily with decoys, ducks, brant, gear, guns, men. The craft shuddered with the impact of the waves, and the spray struck the men with a cold sting, like sheets of hail. The calls of the birds could not be heard above the motor, but a godwit flitted by and the ducks swarmed like black bees against the reddening sun.

Even with a powerful motor, progress was slow. The New Yorker thought of the old-time baymen who crossed in dirtier weather than this, propelled only by their own strength, stout oars, courage, and the strange drive that motivates men who lift "long ducking-guns . . . on bleak, wintry, distant shores."

*Snow geese descending on Chincoteague and buffleheads skimming over Jersey's marshes are as common in fall as sea and shore birds at sunset on Delaware Bay.*

*Flyway researchers have banded redhead, cruising estuary, and European widgeon patrolling beach. Black duck stands in brackish Carolina marsh. New Jersey's Great Swamp (right) is waystop for many geese and ducks.*

*Startled barnacle goose and mallard, both in running takeoff,
contrast sharply with serene flotilla of Canada geese,
preening green-winged teal, and pintails in autumnal garb.*

Scaup are unruffled by offshore chop, and blue-winged teal feed contentedly in shallows. Gunner is setting out goose decoys in iron-gray of morning.

# Chapter Six

# EARLY TIMES AND THE ORIGIN OF LAWS

A wood duck drake flashed through a column of sunlight and disappeared beyond a hanging moss curtain into a dark maze of cypresses and live oaks. An alligator dozed at the edge of the water. On a hill in the distance a big gray sandhill crane balanced on one leg while scratching at some unseen annoyance with the other. Its tufted rump jiggled slightly and its bald red forehead darted down to peck at an insect. A glossy ibis swayed precariously on a branch overhanging the water, as if unbalanced by its scythe of a bill. Below, purple gallinules cackled and plopped about, and an egret trotted splashingly along, shaded by the canopy of its own outstretched wings.

The waters might have been those of the lower Okefenokee Swamp, funnelling through spongy peat banks into the Suwannee River in what is now Florida, or they might have been the flowage of the great Savannah, destined to become a bustling delta port after James Oglethorpe founded Georgia's first city. The precise setting hardly matters since both ecosystems beckoned to wintering masses of mallards, scaup, widgeon, pintails, teal, gadwall, black ducks, ring-necked ducks, blue geese and snows and Canadas. And numbered among the more permanent inhabitants were mottled ducks and fulvous tree ducks and many other species.

In the rushes screening one bank, an almost naked Indian crouched. Before him on a sloping bulge of shore were duck-shaped mounds of mud, a few of them equipped with lifelike heads of bound rush tied on pointed sticks. Similar heads jutted among the grasses that almost concealed a tiny creek to one side of the mud heaps.

A hen mallard dropped into the water and paddled by, head cocked slightly toward the crude decoys. She turned into the little creek, from whose hidden fastness could be heard the chatter and quack of feeding ducks—not yet cognizant that they were imprisoned in a camouflaged weir, like a fish trap but screened above with reeds. Grinning, the hunter closed the mouth of the trap and bent to begin preparations for a feast.

*Oil by George DeForrest Brush, "Indian Hunting Cranes in Florida,"
recalls innocence of age when birds were stalked by bow hunters.*

Far to the north, in marshes and swamps that would some day be
part of Massachusetts or Pennsylvania or Vermont, ducks and geese
fell prey to other Indian hunters armed with spears or bows or nets.
Waterfowl were harvested by the red man for many centuries along the
entire Atlantic Flyway. The primitive methods were clever enough to
be effective. Even today, mud-heap decoys are occasionally used on the
Mississippi shoals and in the northwest by shooters armed with repeaters of the latest design. Although accurate mimicry can be tremendously important in luring some species of wildfowl, snow geese are
drawn to fields by scatterings of ragged white sheets and many diving
ducks are tolled in by "poor man's decoys," plastic bleach jugs painted
black. In modern times, Central American Indians have been observed
killing ducks with three-pronged spears. It is said that an Indian so
armed must be a good stalker and spearman to take four birds in a day,
but it would appear to be no great feat with the less wary species of
ducks early in the season or where hunting pressure is light.

Before the arrival of the white man, when the Atlantic Flyway was
an immense wilderness with mile upon mile seldom trod by a human
foot, when the warning thunder of gunfire was unknown, primitive
methods sufficed.

Surely the same techniques or variations of them were employed by diverse tribes across the entire North American continent. In 1911 anthropologists began to sift the dust of a cave in Nevada for traces of a pre-Columbian culture. Eventually the dig revealed, among other artifacts, decoys fashioned over a thousand years ago by the Tule Eaters, ancestors of the Northern Paiutes. Heads and skins of Canada geese, mergansers, and other species were mounted on forms of tule rush, and canvasback decoys were made of bundled tule alone, rendered lifelike by brownish-red and black paint and the insertion of white feathers. In modern times Ojibway Indians in eastern Canada have been known to attract ducks with still more primitive decoys made from the heavy bark of cork elm—thick, flat, crudely carved silhouettes in feeding attitude, adorned with real duck wings and mounted on a floating slab.

Joel Barber's treatise on decoys reminds us that a French seventeenth-century explorer, the Baron Lahontan, making his way down the Canadian rivers to the head of Lake Champlain, where Vermont now meets Quebec, encountered skillful aboriginal hunters who were using blinds as well as lures. On May 28, 1687, Lahontan penned the following account:

In the beginning of September, I set out in a canow upon several rivers, marshes and pools that disembogue in the Champlain Lake, being accompany'd with thirty or forty of the savages that are very expert in shooting and hunting,

*Canvasback decoys were fashioned over one thousand years ago.*

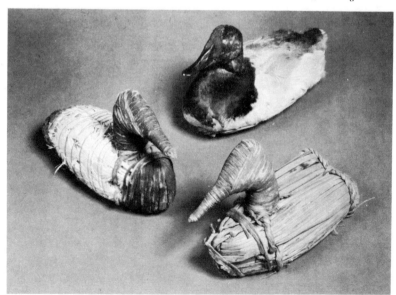

and perfectly well acquainted with the proper places for finding waterfowl. . . . The first spot we took up was upon the side of a marsh or fen, of four or five leagues in circumference: and after we had fitted up our huts, the savages made huts upon water in several places. These huts were made of the branches and leaves of trees and contained three or four men. For a decoy they have the skins of geese, bustards and ducks, dried and stuffed with hay. The two feet being made fast with two nails to a small piece of light plank, which float around the hut. The place being frequented by wonderfull numbers of geese, ducks, bustards, teals and an infinity of other waterfowls—see the stuffed skins swimming with their heads erected as if they were alive. They repair to the same place and so give the savages an opportunity of shooting them either flying or upon the water, after which the savages get into their canows and gather them up.

They have likewise a way of catching them with nets, stretched upon the surface of the water at the entries of the rivers. In a word we eat nothing but waterfowl for fifteen days, after which we resolved to declare war against the Turtle Doves.

Although the Indians of eastern America relied for meat primarily on deer, turkey, and small furred game, their fondness for wildfowl was confirmed by many observers. A youth named James Smith, who was later to fight the British as a colonel during the Revolution and again in the War of 1812, fell captive to France's Indian allies in Pennsylvania in 1755, soon after the eruption of the last and bloodiest of the French and Indian Wars. After being taken to Fort Duquesne (the present site of Pittsburgh), where he saw the Indians burn surviving prisoners taken at Braddock's defeat, he was adopted by a band of roving Mohawks. He remained with them, hunting and wandering through Pennsylvania, Ohio, and Quebec until 1759. Upon reaching Montreal in the summer of that year, he escaped, surrendered to the French, and was eventually released in a prisoner exchange. In a memoir of his captivity, he reminisced about hunting with the Indians in the wilderness around Lake Erie.

He recorded that in March of 1757, south of the lake, "when we came to the Great Pond we stayed there one day to rest ourselves and kill ducks and geese," and a little later in the same area "we remained several days and killed a number of geese, turkeys, ducks, and swans." In November, evidently encamped on the north side of the lake with fifty-three hunters,

we had now great feasting and rejoicing as we had plenty of hominy, venison, and wild fowl. The geese at this time seemed to be preparing to move southward. The Indians present them as holding a great council at this time concerning the weather in order to conclude upon a day that they will all leave at one time the Northern Lakes and wing their way south. When matters are brought to a conclusion and the time appointed that they are to take wing, they say, a great number of expresses are sent off in order to let the different bands know the result of this council, that they may all be in readiness to move at the appointed time. There is a great commotion at this time among the geese. Certain it is that they are led by instinct to act in concert and to move off regularly after their leaders.

The oldsquaw duck, a species whose major autumn and winter habitat is on the Atlantic Flyway, memorializes the Indian attentiveness to wildfowl behavior. It was named by the Crees, who likened its incessant chatter to that of an old woman. (In their own language, they also bestowed on it a number of nicknames which mimicked its calls, and the white traders saluted its tonal range by naming it the "organ duck.")

It was from the Indians, too, that the settlers first learned about the incalculable masses of ducks and geese breeding far to the west, in prairie sloughs and swamps beyond the Mississippi and even beyond the distant Rocky Mountains. After the Louisiana Purchase more than doubled the national domain, exposing great new numbers of nesting birds to the ravages of unenlightened agriculture, it would seem that only Providence saved the waterfowl from going the miserable way of the bison. The white man's farm proved to have a far greater impact than his gun.

The Indian love of wildfowl, and constant quest for meat, had no effect whatever on the population of ducks and geese. Even with more sophisticated weaponry the aborigines would not have depleted the supply. They were too few, too scattered. Like all other predators, these hunters were elements in the ecosystem, but in terms of nature's balance they were far from crucial elements. Neither their presence nor their absence could have tipped that balance as the arrival of white civilization soon did.

Still, contrary to popular notion, early settlement was not characterized by constant game-hoggery but only by an inability to visualize a limit to nature's profusion in the new wilderness. Peter Matthiessen's well-documented book, *Wildlife in America,* makes clear that the colonies founded at Jamestown in 1607 and at Plymouth in 1620 "prevailed because, unlike previous settlements, they managed to live off the land."

Captain John Smith, in his *Map of Virginia,* took pains to describe some twenty birds, and in September, 1621, Edward Winslow of Plymouth wrote that "for fish and fowl, we have great abundance. . . . The country wants only industrious men to employ; for it would grieve your hearts if, as I, you had seen so many miles together by goodly rivers uninhabited."

*John Smith with captors*                    *George Calvert, Lord Baltimore*

The somber Pilgrims of Plymouth were scandalized by the "unseemly appetite for diversion" of Thomas Morton, an Anglican trader who dared to revel about a Maypole at Ma-re Mount, a neighboring settlement which he helped to establish and which inevitably came to be known as Merry Mount. (Nathaniel Hawthorne's story "The Maypole of Merry Mount" is based on his adventures.)

Morton's *New English Canaan,* published in England in 1637, was one of the first books to describe in any detail the animals of North America. He wrote of discovering "pide Ducks, gray Ducks, and black Ducks in great abundance" as well as "Millions of Turtledoves . . . whose fruitfull loade did cause the armes to bend."

No doubt Morton, harassed by rival Puritan fur traders, would have been happier among the sport-loving Calverts of Maryland. Letters written by George and Leonard Calvert in 1633 spoke rapturously of the "infinite" numbers of birds "diversely colored . . . eagles, bitterns, swans, geese, partridge, ducks, red, blue, partly-colored birds, and the like. By which it appears, the country abounds not only with profit but with pleasure. And to say truth there wants nothing for perfecting of this hopeful plantation; but greater numbers of our countrymen to enjoy it."

As a rule, however, colonists went fowling for meat, not sport. It is impossible to say where American duck shooting was first conducted as a food-gathering chore except that the location must certainly have been somewhere on the Atlantic Flyway. As a sport, there is no question that it began in Maryland and Virginia.

Cotton Mather

William Bradford

The New England view of wildlife continued for some time to be an attitude of grim utility, clearly expressed by Cotton Mather's pronouncement that "what is not useful is vicious." If Mather was somewhat less familiar with wildlife than with witches, he nevertheless fancied himself a naturalist, and one of the revelations in his *Philosophical Transactions* was that wild pigeons, in their annual migrations, betook themselves to an "undiscovered Satellite, accompanying the Earth at a near Distance." The doomed passenger pigeons were already being slaughtered in great numbers for food. In their descent upon the Plymouth crops in 1643, a mystic might easily have discerned nature's ironic vengeance: the threat of famine. For the pigeons, it was at best a pyrrhic victory. A similar onslaught of the birds in 1648, when crops failed, actually prevented a famine.

Such episodes seemed to vindicate Mather's implied dictum that the only good wild creature was a dead wild creature, killed for food.

The earliest game regulations were intended either to encourage the killing of animals for food and fur or to encourage the extermination of unwanted species. The bounty system was established by the Massachusetts Bay Company in 1630, with the inducement of a penny per wolf. The meagerness of the payment probably indicates the astounding abundance of a species that was to disappear from New England by the mid-nineteenth century. Other settlements soon joined in the campaign of extermination. William Penn offered bounties in 1683 to rid his colony of wolves thought to be preying on livestock. In 1697 New Jersey offered twenty shillings per wolf to any "Christian," and half that sum

*67*

to any Negro or Indian. (One wonders if perhaps the Negroes and Indians were such superior hunters that they brought in twice as many wolf skins.)

New York paid half a dozen wolf bounties as late as 1897, and only now are conservationists and state game departments beginning to overcome the deep-seated prejudices of American farmers and ranchers regarding predatory animals. The bounty system was not established to benefit waterfowl, upland birds, and the other natural prey of the carnivores but to subsidize producers of domestic chickens and lambs and such. Coincidentally and to a minor degree, the bounties (together with the general affection for fox and raccoon fur and general dislike of snapping turtles) may have increased the wildfowl population for short periods. But if man had not placed a heavy thumb on the scales of nature, such measures would have been entirely irrelevant, for a healthy habitat—undrained, unfilled, uncrowded, unpolluted—balances itself, with sufficient predators to cull the aged or ill or unfit superfluity of prey, and a sufficient superfluity of prey to maintain the habitat-purifying predators. In this way natural selection perpetuates a healthy population.

Since man did place his thumb on the scales, some redress is demanded, but the elimination of all predators only makes matters worse. First, it leads to a genetic weakening of game species by removing an element of natural selection. Second, it further aggravates the imbalance by eliminating an effective control on other species which can become far more undesirable—for example, the small but indefatigable rodents which devour needed food and may transmit parasitic scourges.

The earliest regulations, then, were misbegotten, but this is not to say that all of the settlers were of a mind with Cotton Mather. William Bradford, governor of Plymouth Colony and a strong advocate of the need to live off the land (and therefore to keep the habitat productive) foresaw a waterfowl decline. At first such pessimism went unheeded but in 1710, less than a century later, Massachusetts banned the use of boats, sailing canoes, or camouflaged canoes in the taking of waterfowl. New York had already acknowledged the impact of civilization on game, having enacted in 1708 a law providing closed seasons in some counties on the heath hen, grouse, quail, and turkey, but many more years passed before the colony sought to alleviate hunting pressure on migratory birds.

New York, in fact, no longer was a colony when the effect of unbridled meat-hunting became worrisome, for it was not until 1791 that a law was passed to halt the taking of woodcock during the breeding season in

*George Catlin's 1857 painting, "Shooting Flamingos," illustrates
19th-century lack of restraint in harvesting dwindling game.*

the bucolic boroughs of Manhattan, Brooklyn, and Queens. This was
probably the first such law on the flyway or, for that matter, anywhere
in the country.

Maryland prohibited firelighting—taking waterfowl and other game
at night with a gunning light—in 1730, as did North Carolina in 1777
and Virginia in 1792. Virginia simultaneously outlawed the cannon-
bored punt gun. However, such laws were generally not taken very
seriously until the twentieth century, and the reason for this was not
simply a lack of manpower to enforce the restrictions. It was a matter
of attitude, both on the part of the shooters and the legislators who
passed the laws.

This attitude was rooted partly in America's revolutionary dislike of
authority and partly in the necessity of living off the land in a new,
untamed country. Birds and animals were primarily hunted not for
sport but for food or profit. In some areas hunting continued to be more
often a matter of subsistence or livelihood than of sport until almost
half of the twentieth century had passed. This is why Maryland, so
quick to perceive the evil of firelighting, was among the last states to

*Wood engraving shows use of battery box on Potomac in 1850.*

permit wildfowling in the spring, the sale and export of wild game, unlicensed hunting by state residents, and the killing of nongame birds. (Before market gunning was abolished, robins and blackbirds sold for twenty-five cents to seventy-five cents a dozen, and bird pies remained sufficiently popular in the Roaring Twenties and the Hungry Thirties so that wardens were able to confiscate several hundred robins during a single Louisiana raid on black-market gunners.)

Eventually, some prohibitions evidently were enacted on the basis of the chivalrous American sense of sportsmanship—part of the English heritage. But until hunting became a sport, sportsmanship was irrelevant.

Aside from measures taken to eradicate animals considered to be vermin, all of the early regulations seem to have been enacted either because the decline of a particular food species in a particular place (the woodcock in the coverts of urban New York) became inescapably noticeable, or because a gunning technique interfered with some other technique of greater economic importance. For example, it was widely believed that firelighting and punt gunning (often used in combination) quickly caused waterfowl to avoid the easily accessible, highly productive areas where a most popular hunting device was the sinkbox—or battery box, as it was then called.

This belief was probably quite sound. Observers agreed that in regions where firelighting was common, ducks soon began to fly away in fright each evening when residents lit the lamps in their homes. Experienced hunters know that the warier species of wildfowl quickly learn to associate the sound of gunfire with danger, and flare away from it. If this is true with respect to guns of ordinary size, using charges of reasonable magnitude, it must also apply to the much louder boom of a punt gun which can be heard at far greater distance. Where gunning lights and punt guns were banished, they were outlawed not for devastating whole flocks of birds at a single shot but for *failing* to kill every bird in a flock. Each escapee forevermore feared lights and gunfire, and communicated this fear to its flying companions. Thus was an area's productivity lowered, to the chagrin of the market gunners who fired from battery boxes and soon also began to rent these submersible blinds to a new breed known reverently as "the wealthy sportsman."

In some areas, so many sinkboxes pocked the waters that it was feared these devices, too, might lower an area's productivity. But restraints of any kind were likely to arouse the moral and patriotic indignation of this young republic's Sons of Liberty. Legislative prohibitions sometimes incurred the approximate degree of outrage to be expected in the event of renewed taxation without representation. New York banned sinkboxes in 1839. Gunners soon began to defy the law, wearing masks to prevent identification. After a little more time the masks were discarded on the logical assumption that no man wishes to be shot, therefore no man is foolhardy enough to attempt the enforcement of such a tyrannous law or to inform on violators.

The legislatures (or, in later years, the game commissions) of other states eventually found the courage to abolish sinkboxes—Ohio in 1852, New Jersey in 1879, Michigan in 1897, Maryland and North Carolina in 1935.

Theoretically, the battery box is illegal everywhere now, but how does one classify the "curtain blinds" of North Carolina's Pamlico Sound? Perfected on Ocracoke Island and now maintained by guides along other stretches of the Outer Banks, these are concrete pit blinds sunk in shoal water, with wooden extensions rising just above water level, and with wooden wings and canvas curtains to deter waves from sloshing in and perhaps to aid in camouflage as well. Apart from the fact that a curtain blind has to be pumped out each morning and is immobile, it is very like the sit-up style of sinkbox. When a hunter sits inside, his head is just a little above water level. This remains quite as legal as the use of a pit blind on land. There is no reason to abolish such contrivances, for they can hardly be devastating to wildfowl if the legal seasons, bag limits, and other regulations are observed—as they are in the main, since they are now rigidly enforced.

Also in legal use are various camouflaged sneakboats and "layout" boats. The gunner lies supine in a layout boat, just as market shooters did in the old "lay-down" sinkboxes. The craft rides almost as low in the water as a battery, and its frame is covered and decked over with fiberglass canoe cloth which keeps the water out. The layout boat differs from the battery box in two major respects, however: it floats a little above the water, so it is federally legal, and it is reasonably seaworthy if handled with care. It can be argued that the battery box fell into disuse not so much as a result of legal reform but primarily because it was an unsafe craft, too easily swamped when the water became a trifle rough.

Battery boxes, night lights, punt guns—these and other clever if deplorable devices of a profligate age were all developed along the Atlantic Flyway and exerted a significant influence on nineteenth-century gunning practices. They merit closer scrutiny, but not yet. For even before they had been developed, the specter of a wildfowl decline loomed over the Atlantic Flyway. And no one was sufficiently perceptive to see it.

*American faith in unending natural bounty permeates old sporting prints.*

*Long Island is believed to be scene of 1852 print, "Wild Duck Shooting."*

# Chapter Seven

# EARLY TRAGEDIES AND THE BEGINNING OF REFORMS

Assuming that sanity overtakes the rushing multiplication of the human species and that the gradual but unremitting destruction of habitat can be halted, the trove of wildlife on the Atlantic Flyway can remain copious for many centuries to come. In view of society's belated ecological awareness, faddish though it may be to some degree, there is reason for measured optimism. It was not always so, and the temptation to find simple explanations for complex catastrophes led to the charge that hunters had devastated the waterfowl population. The truth is that hunting for sport has never endangered any American species. Even meat-hunting, whether for subsistence or for the market, has rarely devastated any species and would not have presented any threat at all to wildlife populations if it had not been accompanied by a massive reduction in habitat. In *The Outlaw Gunner,* a book dealing with nineteenth and twentieth century market shooting, Harry M. Walsh has pointed out that an area three times the size of all New England has been removed from use as prime wildfowl-breeding area by pollution and drainage.

The near-extinction of the whooping crane frequently has been exploited as a condemnation of hunting, and it is true that shooting—both for food and for sport—accelerated the decline of this bird which was so rare to begin with. However, the course of evolution is beset with casualties, and this species had failed to adapt well to conditions having nothing to do with the human predator. As Peter Matthiessen (and other authorities) stated: "Probably there were no more than a few thousand of these birds in all America at the time of the first records, and it must be assumed that the species, declining slowly over thousands of years, was then only a few centuries from extinction."

Admittedly, there have been several fairly convincing exceptions to the rule that food hunting was not the great executioner—the Eskimo curlew, the heath hen, the passenger pigeon come to mind. But even in such cases, other factors were usually critical. Historical research by the Fish and Wildlife Service indicates that loss of habitat was important in the decline of the heath hen and passenger pigeon. The pigeon, moreover, had much in common with the persecuted predators; because it was potentially destructive of crops, it was sometimes considered vermin to be slaughtered for the greater prosperity of man.

It was not commercial food hunting but another type of market killing that threatened many birds of the Atlantic Flyway. For the millinery trade, established early, was as devastating to the creatures of the air as fur traders were to the creatures of the land. The whooping crane was killed for feathers as well as flesh. Among the species ravaged by plume hunters were the common tern, great white heron, roseate spoonbill, reddish egret, flamingo, trumpeter swan, brown pelican, wood duck, and Carolina parakeet. Most of these species were belatedly rescued from the edge of oblivion. The extinction of the Carolina parakeet was perhaps inevitable, owing to its loss of forest habitat and its penchant for destroying fruit orchards.

Salvation for the milliners' victims finally came towards the close of the nineteenth century when protective legislation was passed, largely through the lobbying and proselytizing efforts of the vigorous young National Audubon Society. The formation of the society had been chiefly promoted by the outdoor magazine *Forest and Stream* under the guidance of its editor, George Bird Grinnell, who was also an original member of the American Ornithologists' Union, cofounder with Theodore Roosevelt of the Boone and Crockett Club, and an indefatigable wildfowler.

Ironically, within a few years the Audubon Society became so vehemently protective of all birds, including game species, that mention of its very name was avoided in the pages of Grinnell's magazine. To end all harvesting of game birds is to doom them by overburdening a depleted and still shrinking habitat, but there were (and are) foes of hunting who refuse to concede this inevitability.

*Whooping crane was nearly exterminated by meat marketers and feather merchants.*

*John James Audubon*      *Theodore Roosevelt*      *George Bird Grinnell*

*These hunters and other outdoor enthusiasts awakened popular interest in wildlife study and conservation. Grinnell, known as "father of American conservation," edited finest sportsmen's journal of 1880s and '90s. Audubon's portrait (painted by his son) shows him with fowling piece, and Roosevelt poses with big-game rifle.*

If Grinnell were alive today, he might be even more dismayed by the views of some of the organization's members, but there is no denying that the Audubon Society has been instrumental in protecting many threatened forms of wildlife and awakening the public to the need for conservation. Its success in fighting the plume hunters probably came just in time to save a number of species from extinction, for as late as 1886 the American Ornithologists' Union published the revelation that five million birds were being killed annually in this country just for the adornment of ladies' hats.

The struggle to enforce the new protective laws and halt this carnage was greatly impeded by the lingering Matherian attitude that "what is not useful is vicious," that all creatures were put on earth by God exclusively for man's consumption. It is hardly surprising that in 1905 a warden named Guy Bradly was murdered by an enraged Florida plume hunter on Oyster Key. Nor is it really surprising that a local jury, honoring the tradition that game is expendable, set the killer free.

The tradition was a venerable one. A farsighted colonist named John Josselyn fretted in 1672 that the passenger pigeons "of late . . . are much diminished, the English taking them with nets," but sixty-five years later, when the great flocks of the East were already disappearing, Boston cooks were buying them at six for a penny.

By that time, the great feather trade had begun and other species were being decimated. Perhaps the "pide Ducks" recorded by Thomas Morton in 1637 were buffleheads or oldsquaws or eiders, perhaps they were brant or ring-necked ducks, golden-eyes or scaup. But he may well

*Painting by Louis Fuertes depicts Labrador ducks on Arctic coast before millinery trade hastened their extinction.*

have been referring to the handsome little black and white Labrador duck, frequently called the pied or sand-shoal duck. Although no one has discovered the full reasons for the demise of this hardy species, it is believed that the heavily contributing factors included the many "feather voyages" made to the nesting grounds on the southern coast of Labrador by New England ships in the 1750s.

The colonists also sought the feathers and down of the eider and other sea ducks breeding in Labrador. Greater numbers and a wider range enabled these species to survive, yet the voyages ceased for lack of victims after hardly more than a decade. And even after that, the plundering continued, for the Indians and fishermen of Labrador seemed unable to satiate their craving for duck eggs.

Apparently the Labrador duck never quite recovered from these first onslaughts. The species, which had formerly wintered from Nova Scotia down the Atlantic coast as far as Chesapeake Bay, was uncommon by the first quarter of the nineteenth century. John James Audubon, though he wrote that the Labrador duck "at times enters the Delaware River, in Pennsylvania, and ascends . . . at least as far as Philadelphia," was unable to find a live one. Wishing to describe it, the great naturalist-artist-huntsman was forced to rely on a pair of birds shot by Daniel Webster off Martha's Vineyard and later ensconced in the Smithsonian Institution.

Audubon and his fellow ornithologist Alexander Wilson would have been appalled to know that their accounts were in large part responsible for the later nest-robbing and egg-collecting craze which so threatened both game and song birds that the practice had to be curtailed by legislation—again sponsored by the Audubon Society. But even if the importance of bird identification had been recognized in the mid-nineteenth century, sensible game management was still in the future. Men continued to kill the few remaining Labrador ducks. One was said to have been taken near Elmira, New York, in 1878. But the last authenticated specimen was shot three years earlier, on December 12, 1875, by a Long Island gunner.

In spite of the utilitarian and wasteful attitude toward game, there were some auspicious omens in addition to the early legislative gropings thus far cited. Not surprisingly, the elusive deer rather than the sky-veiling clouds of birds were the first beneficiaries of new measures. The Massachusetts Colony proclaimed a closed season on deer in 1694, and the first game law that might be called federal was a closed season on deer in all colonies but Georgia in 1776. Regrettably, some of the later laws—such as one protecting Massachusetts pigeon netters from molestation in 1848—were actually detrimental to birds.

But it was also Massachusetts that established the first warden system in 1739. New York emulated her sister colony two years later. It was Massachusetts, again, that established the first closed season on larks and robins in 1818 and on heath hens in 1831.

Three years before this recognition of the heath hens' plight, President John Quincy Adams had set aside the Naval Live Oaks Plantation in northern Florida to conserve the live-oak trees used in building sailing vessels. This appears to have been the first attempt to conserve any resource other than animal life, and of course the preservation of such a woodland amounted to the preservation of habitat not only for migratory birds but for other wild creatures. The President's little oak plantation is now contained in the Gulf Islands National Seashore which stretches across more than one hundred and fifty miles of Florida and Mississippi and is one of the latest National Park tracts to be snatched from the game-killing tentacles of "development."

That most of the first steps toward conservation were taken somewhere on the Atlantic Flyway does not indicate any clairvoyance on the part of eastern Americans. It was merely that the East was so heavily and quickly settled as to render inescapably obvious the signs of environmental rape. Even so, game protection invariably came first to states where it would interfere with no one's livelihood. Bird pies were popular in the antebellum South, and due to become more so in the hungry years of Reconstruction. Thus, full protection of nongame birds arrived first in northern states—Vermont in 1851 and Massachusetts in 1855. A measure of protection was granted to pigeons in Michigan in 1869, but the first eastern state to follow suit was Pennsylvania, nine years later.

There were, however, some frugal Yankees along the Atlantic Flyway, men who thought clearly about what was happening to wildfowl and other game. In some quarters, such matters were taken seriously by 1844. The New York Association for the Protection of Game, founded in that year, was the first American conservation group.

Two years later Rhode Island passed the first law against spring shooting of wood ducks, black ducks, woodcock, and snipe. Though the regulation was soon lifted by popular demand, men were gradually accepting the revolutionary idea of saving "seed stock" to ensure a future supply of game.

This, in turn, led to the acceptance of a periodic need to relieve waterfowl from hunting pressure even in the Atlantic Flyway states that clamored most vociferously for the flesh of ducks and geese. Acceptance of such periodic relief, in the form of "rest days," was facilitated by the American regard for the Sabbath. Thus, in 1872, Maryland, a leading wildfowling state, passed the first law setting aside rest days for waterfowl.

A year previously, the Federal Commission of Fish and Fisheries had been established, and before the decade was over Arkansas had prohibited market gunning, Florida had begun protecting the eggs and young of plume birds, and the first game commissions had been formed by two states on coastal flyways—New Hampshire and California— while Iowa, on the Mississippi Flyway, had set the first bag limits on game birds.

In the 1890s, the modern system of resident and nonresident licenses was inaugurated in a number of states, and the Supreme Court (in Geer vs. Connecticut) decreed game to be the property of the state rather than the landowner. In 1900, the Lacey Bill was passed, forbidding the

*Aquatint shows bird shooters and anglers at High Falls of New York's*
*West Canada Creek in 1835, before onslaught of urbanization and pollution.*

*Earliest game law was closed season on Massachusetts deer.*

importation of foreign animals or birds without a permit, and Congress at last forbade interstate traffic in creatures killed in violation of state laws. The following year, New York, New Jersey, West Virginia, and all of New England except Rhode Island responded to an alarm issued by the recently founded Biological Survey by proclaiming closed terms on the wood duck. Then, at Pelican Island, Florida, in 1903 the first federal wildlife refuge was established.

There were setbacks, certainly; in 1906, Congress rejected appropriations for continued investigations by President Theodore Roosevelt's Natural Resources Committee. But the conservationists were not beaten. Two years later Roosevelt set up the National Conservation Commission under Gifford Pinchot; in 1910 New York prohibited the sale of wild-game meat or plumage; and in 1913 the Weeks-McClean Act awarded responsibility for migratory game birds to the Biological Survey. Thus, thanks to the accepted view that interstate matters are federally controlled, government agencies could now assume a proprietary role in behalf of migratory waterfowl and shorebirds, building a foundation for ultimate protection on a national scale.

Unfortunately, the treatment accorded to habitat during the three centuries of slow enlightenment just summarized did not keep pace with the treatment of birds and animals. This was true in spite of the fact that in the beginning the land fared better than the fowl which depended on it. According to a 1968 publication of the Maine Department of Fisheries and Game, "tidal marshes were the most valuable asset to coastal communities in colonial times because of the hay crop, [and] for several generations the marshes lay undisturbed." But while treatment of the game gradually improved, treatment of the habitat worsened "until twentieth century earth-moving machines began their systematic destruction of tidal marshes along the Atlantic coast.

"Inventories of coastal wetlands, made by the Bureau of Sport Fisheries and Wildlife, reveal that in New Jersey, for example, 24,609 acres of the state's coastal wetlands, nearly ten percent of the total, were lost between 1954 and 1964."

A ten percent loss in a single decade was by no means unique to New Jersey. During the eighteenth and nineteenth centuries, many inland farmers in Maine owned sizable pieces of salt marsh, where hay was cut and stacked just above the level of the highest full-moon tides so that it could be sledded home when winter came. Old-timers in Maine can still recall the marsh shoes worn by horses, but the beginning of the twentieth century marked a sudden development of many barrier beaches rimming those wetlands. Trolley and railway lines proliferated, bringing with them swarms of summer visitors who preferred to replace bleak marshland with urban accommodations reminiscent of those at British seaside resorts. The new mobility of tourism, like the new mobility of the industrial revolution, accelerated the corruption of the wetlands.

The marshes were darkened not by gunsmoke but by chimney and factory smoke and the billowing black smokestacks of harbor-building shippers. By the eighteenth century, overcrowded urbanization had become a problem in Old New York, with the concomitant horror of rampant pauperism—the threat of hunger, the alms house, the debtors' prison. Free or cheap game meat might tide a man over, and exploitation of the land might give his family security.

In 1810, the population of Connecticut was 261,942, that of Delaware 72,674, that of South Carolina 415,115. All the people in New York and New Jersey were fewer in number than the population of several towns within those states a few generations later, but in terms of unspoiled habitat the Atlantic Flyway was already overcrowded with human beings and their bustlings and edifices by 1810. Without including Florida, where no census had yet been taken, there were at least six million people along the Atlantic Coast. And the figure of six million was paltry by comparison with the population figures at the dawn of the next century.

*Brown pelicans have been menaced by plume hunting, destruction of habitat, water pollution, and pesticides.*

In *The Treasury of Hunting,* by the late Larry Koller, one finds this most revealing passage concerning the human confiscation of wildfowl habitat:

The depletion of native game can be blamed in large part on the market gunners; not, however, the reduction of wild fowl. It is certainly true that colonial America teemed with ducks and geese, that the migrating flocks would appear in great clouds, that the flushing of birds from the water sounded like thunder. How many birds such a profusion represented, no one can say for sure. Probably it was somewhere in the billions.

The market gunner took his share, no doubt of that. He shot millions of birds each year, relentlessly, with shotguns, swivel guns, and punt guns, by baiting them into traps, by hitting them in both the spring and fall migrations. It was slaughter, but it is probably safe to say that it barely harvested the annual crop of birds. For with optimum nesting and feeding conditions, the broods of young were enormous.

As long as the birds were left alone on their nesting grounds, the supply of ducks was unlimited. But as early as 1849, when, with the passage of the first Swamp Act, some 70,000,000 acres of northern breeding grounds were drained, the decline of the duck had begun. The cutting of hardwood timber in the North Central states decimated the tree-nesting species, among them the beautiful wood duck. The final blow came in the early years of the twentieth century with the draining of 100,000,000 acres of American wetland nesting areas to make way for more wheat planting. Wetlands in Alberta and Saskatchewan also were converted to wheat growing. The loss in ducks and geese through these drainage programs is incalculable. Furthermore, the losses continue. Between 1943 and 1961, more than 1,000,000 acres of wetland were drained in the North Central states, most of it prime nesting area for wild fowl.

And Koller did not even mention the dust bowl years of the Great Depression, when wildfowl-devastating drought visited the land, and there was insufficient money to save people, much less ducks and geese.

Yet, during this same bleak period, a new conservation movement was growing, new laws and continent-wide programs were established, and sportsmen were beginning to take the need for wildfowl protection and maintenance into their own hands. Hindsight seems to indicate that today's ecological awakening was as inevitable as it was tardy, exemplifying the frequently reiterated observation that Americans are congenitally crisis-oriented and also congenitally optimistic: If tragedy comes, can reform be far behind?

Moreover, the "bleak period" was in some respects far from bleak. It was the period of the professional watermen and the golden age of wildfowling for sport. All was not tragic when, in defiance of man's mightiest mistakes, clouds of ducks and geese did veil the skies.

# Chapter Eight

# REFLECTIONS ON DOGS, DECOYS, TETHERS, AND CALLS

There is something inexplicably appealing in the fancy that water dogs spring from stock employed, bred, trained, perfected by old-time professional gunners. Perhaps the desire to believe the fable is unconsciously founded on the realization that those gunners were so adept, at their trade. Any breed used and developed by such men must be talented, indeed. Or perhaps the appeal is merely symptomatic of democracy romanticized—America's apotheosis of the proletariat. The truth is that with the exception of dog packs used by primitive hunters, and breeds employed for hauling or herding, "working" dogs have almost invariably been luxuries, cultivated and cherished by aristocratic sportsmen.

Few such animals have been truly essential to the purpose for which they were employed. Without a pointing dog, it is true, the quail hunter might almost as well stay at home, but a typical wildfowler of the seventeenth or eighteenth century would have considered a retrieving dog as no better than an extra mouth to feed since the animal could not have found or held game within gun range. Regrettably, numerous records and accounts show that until quite late in the nineteenth century it was rare for a market gunner to bother chasing down crippled birds. In some regions, it was equally rare for a sportsman or sportsman's guide to do so.

Since canine companionship was a luxury of the wealthy, the development of almost all sporting breeds took place in Europe. Bird dogs were portrayed in a few of the English aquatints that were circulated in America, and the Boston artist John Penniman copied several of these aquatints at some time between 1805 and 1820, but an earlier and original American picture is of greater historical interest. It is Benjamin West's portrait of Thomas Mifflin, painted circa 1758. It shows a happy fourteen-year-old boy posing with his fowling piece and several ducks he has bagged. Barely discernible in the background is a large retrieving dog of uncertain lineage, swimming across a pond. The young man in

*Benjamin West's portrait
of Thomas Mifflin, painted
in 1750s, may be first
American work of art
to incorporate theme of
wildfowling as sport.*

the portrait was destined to serve as an aide to General Washington and later to become Governor of Pennsylvania. Significantly, the Mifflins were well-to-do; they could afford to commission portraits and to hunt in the European fashion.

Also significantly, only three major breeds of hunting dogs have been truly American in origin and, as might be expected, all three are natives of the Atlantic Flyway. They are the Chesapeake Bay retriever, the American water spaniel, and the black-and-tan coonhound. The oldest of the three breeds is the black-and-tan, according to John R. Falk's *The Practical Hunter's Dog Book,* which notes that this hound's origin is veiled in conjecture but "the general consensus holds that the black-and-tan evolved from the Talbot hound, an ancient breed, now extinct, that was brought from England into Virginia by some of that state's first settlers."

The statement reminds one that the Virginia and Maryland colonists were more favorably disposed toward "frivolous diversion" than were the Puritans of the North. A predominantly black-and-tan strain that sprang from the Talbot stock was dubbed the "Virginia black-and-tan." Almost from the beginning, the black-and-tan was prized not only as a coonhound but for the more aristocratic sport of fox hunting, and the breed also contributed to the development of the Chesapeake Bay retriever, second-oldest American sporting breed.

Another painter, the Irish-American sportsman Hugh Newell, portrayed himself and two wildfowling companions in an 1856 oil entitled *Duck Shooting, Susquehanna Flats.* The scene is set near Havre de Grace, Maryland, on a balmy autumn day. Eight ducks—mostly redheads though one or two may represent canvasbacks—lie on the bank. One of the hunters holds a ninth as he accepts a tenth from a retriever coming ashore. The dog appears to be a water spaniel of some kind. (The date of the painting is a bit early for an Irish water spaniel in Maryland and much too early for an American water spaniel, but the

*Hugh Newell's 1856 canvas depicts use of retrievers on Susquehanna Flats.*

dog certainly looks like a forebear of the American variety.) Sitting on the bank and contemplating two of the hunters' ducks is a larger dog of somewhat ambiguous conformation and color. The animal may well represent a blend of the Newfoundlands and particolored local breeds that played a part in establishing the Chesapeake retriever's ultimate form.

The various accounts of the Chesapeake's evolution agree on several points, beginning with the breed's Newfoundland ancestry. The Newfoundland settlers of the seventeenth century nurtured a big, black, thick-coated breed known as the St. John's Newfoundland. An even larger, white, thick-coated breed called the Great Pyrenees was brought over in 1662 by Basque fishermen who founded a Newfoundland station at Rougnoust. Crosses of the two breeds probably established the foundation stock of the standard Newfoundland. In 1807 a pair of Newfoundland pups—a dingy red male named Sailor and a black female called Canton—were put aboard a vessel bound for England. Either the dogs were rescued from a shipwreck, as some accounts indicate, or were transferred to another vessel headed for Baltimore. Sailor may have been given to John Mercer of West River, Maryland, and Canton to Dr. James Stuart of Sparrows Point, Maryland, or both may have gone to another Marylander named George Law. It is not even known whether Canton had a litter by Sailor. What is known is that both dogs were crossbred with the black-and-tan coonhound. The crossings were reported by Ferdinand Latrobe, former mayor of Baltimore and for most

of his adult life a member of the Carroll's Island Club, a sportsmen's group which played a large part in the development of the Chesapeake.

Historians believe that the black-and-tan infusion produced the breed's yellow eyes and lightened its color to the present dead-grass hue. More important, it is said to have contributed the strong, heavy tail, which is an aid in swimming, and to have greatly improved the Chesapeake's nose and stamina. Further modifications resulted from additional crosses as time went on; one cannot be certain of the breeds involved but they may well have included flat- and curly-coated retrievers and the Irish water spaniel or a similar breed.

John Falk and other authorities have averred that the modern Chesapeake "is beyond question the finest water working retriever in the world. For rugged power, endurance, and skill under the most trying conditions, no other water working retriever can hold a candle to him." The dog's marvelous aquatic abilities led to an early but persistent myth, literally believed by many innocents, that the original breeding stock was the result of a mating between a female retriever and an otter.

In view of such legends, it seems peculiar that the Chesapeake has traditionally been far less popular than the Labrador retriever and the golden retriever. But the Chesapeake is a somewhat flawed work of art. An animal standing more than two feet high at the shoulder and weighing seventy pounds is a bit too much dog for most people, especially in this era of cramped urban quarters when habitat for men and dogs has shriveled together with that of wildfowl. Moreover, the breed's thick double coat is oily—an advantage in cold waters but hardly endearing indoors. Finally, the breed's magnificence in the wetlands is absent in the uplands, where the Lab and the golden can be put to supplemental hunting use. It may be that the Chesapeake Bay retriever is a living relic of waterfowling's golden age, never to regain ascendancy.

The Labrador retriever entered upon the scene just a little later than the Chesapeake and, in spite of the breed name, was developed abroad even though the St. John's Newfoundland was again a chief progenitor. The St. John's, like the standard Newfoundland and the Great Pyrenees, was a fine water dog and evidently one of the few genuine workers. All of those breeds were used occasionally to retrieve waterfowl but primarily to fetch ropes from ship to shore or boat to boat. It should be added that they proved valuable in retrieving something other than ducks and ropes, for they also helped to rescue men from the sea.

Sometime during the first quarter of the nineteenth century, St. John's Newfoundlands were brought to England at the behest of the Second Earl of Malmesbury. Word of the Earl's fine swimming dogs soon reached other aristocratic sportsmen, and the breed's reputation was enhanced by the approval of such famous hunters as Colonel Peter Hawker, who wrote that the Newfoundland was "extremely quick running, swimming and fighting" and that "in finding wounded game there is not a living equal in the canine race."

During ensuing decades, English breeders crossed these dogs with others, probably including flat- and curly-coated retrievers, until a standard Lab finally emerged in the late nineteenth century and was recognized by the English Kennel Club in 1903. Shortly afterward, Labrador retrievers began appearing up and down the Atlantic Flyway, and noted American breeders—J. F. Carlisle, Averell Harriman— did much to bring a splendid new water dog to the attention of sportsmen in this country. It was on the Atlantic Flyway, of course, that Labrador field trials began; the first was held at Chester, New York, in 1931. Since then, the Lab has dominated other breeds both in terms of popularity and of retriever-trial victories. The only three multiple winners of the National Retriever Trial have all been Labs. One of them was Nilo's King Buck, owned by John Olin, another famous breeder. King Buck achieved a unique kind of immortality when wild-life artist Maynard Reece's portrait of the big black retriever became the design for the 1959 Federal Migratory Bird Hunting stamp. Some of the owners of yellow or chocolate Labs may have yearned for a different color scheme on that stamp, but the black Lab has undeniably become the symbol of retrieverdom in America.

The average Lab is almost as big as a Chesapeake but is so sedate and affectionate (and relatively odorless) as to fit in perfectly with the desires of today's pampered wildfowler, who may be an accredited member of the middle class but retains the outlook of the aristocratic sportsman. Then there is that upland ability.

Pretty much the same attributes can be claimed for the golden retriever, whose handsome coat is seen by the sentimentally inclined as an outer symbol of the animal's temperament and skill. Perhaps it was also a sentimental disposition on the part of wildfowlers that accounted for wide acceptance of The Great Golden Hoax—a story to the effect that the English sportsman Lord Tweedmouth acquired the original breeding stock from a Russian circus trainer, whose act he demolished in 1860 by purchasing all eight of the Russian's dogs at an exorbitant price. Allegedly, the dogs were known as Russian Trackers

*Among water dogs, golden retriever ranks with Lab in universal esteem.*

and were renowned back on the Steppes for performing herding duties all winter long and without supervision, while their masters kept their feet near the stove.

Lord Tweedmouth did originate the stock, but with a single yellow pup named Nous from an otherwise black litter of wavy-coated retrievers. Nous was bred to a representative of the Scottish clan known as Tweed water spaniels. From this union there came dogs that were crossed back and subsequently linebred and then outcrossed with Labs and wavy- and flat-coated retrievers, and in 1913 the Kennel Club of England registered a golden retriever. A bit earlier than that, some former officers of the British army had brought the first goldens into Canada, and from there they reached the United States shortly before World War I.

Like the Lab, a golden retriever is both tractable and talented in a duck boat or blind or in the uplands. These cursory breed histories scarcely convey the esteem in which duck hunters have always held their dogs or the pride with which a man watches his canine lieutenant accomplish a difficult retrieve.

It is a rather melancholy and puzzling fact, though, that the smaller, equally amenable and almost as strong American water spaniel experienced only a brief flare of affection among this country's wildfowlers. Technically a true spaniel rather than a retriever, this little dog has a long tail, a relatively rangy body, a wavy or curly coat of liver or dark brown, and an eternally alert, cheerful, eager expression that is not at all hypocritical.

Records of ancestry are scanty, but the breeding probably combined bloodlines of the English and Irish water spaniels and the curly-coated retriever, all of which were being used on the Atlantic Flyway by the

*Irish water spaniel and other curly-coated retrievers*
*were frequently pictured in 19th-century art.*

1880s, when the American water spaniel seems to have emerged. Although no sponsoring club was formed until the 1930s, and recognition by the American Kennel Club did not come until 1940, the breed had long since won the esteem of some discriminating gunners in the northeast and then in the Midwest. There are written references to the American water spaniel, and turn-of-the-century drawings and paintings showing dogs that conform at least vaguely to the standard for that breed.

Still earlier paintings depict Irish water spaniels, their tightly curled livery revealing a heritage of Portuguese water dogs and the South Country water spaniels of Ireland. Even more striking is a resemblance to the poodle. Some authorities believe that the Portuguese water dog was a forebear of the poodle. Some also believe that the poodle itself was a forebear of the Irish water spaniel. Quite possibly. The resemblance, of course, is not to the effete city poodle with the vulgarly fey haircut favored by bench-show dowagers, but to the fully insulated poodle that won French—and international—accolades as a retriever.

Occasionally, this and other breeds not only retrieved ducks and geese but enticed them into gun range for subsequent retrieval. For a brief period here and abroad, sportsmen trained such dogs as tollers. The animals pranced and cavorted near the blind, usually on a bank fronting a wide expanse of water, to tantalize birds. Ducks and geese are extremely inquisitive creatures, and when they are on the water they do not fear four-footed terrestrial predators as they fear man. They seem to understand that a fox or wolf or dog—or even a raccoon—cannot swim out and overtake them. If an enemy were to attempt it, they could simply rise from the water before the completion of the approach. Knowing this, and being inquisitive, they will often paddle directly toward an animal displaying itself on shore. When they are close enough, a sportsman need only rise in his blind to flush them and make his shot as they take off.

Yet dogs are rarely used as tollers any more. Unless a man has a strong compulsion to show off, he has no reason to train his dog as a vaudevillian. Decoys, a properly situated blind or concealed boat, and restrained, infrequent calling are all that is needed. It is true that when the wind is wrong ducks sometimes set down on the far side of the decoys, just out of range when and if they finally rise again. But there is an often successful way of correcting this situation without a tolling dog. The method was learned long ago from the Indians, who also used it to attract caribou in the Northeast, pronghorn antelope in the West. The Indian hunter, keeping himself concealed, waved a red or white ''decoy flag'' which he held up by hand or tied to a pole. On Long Island and on the Eastern Shore, more than one gunner this season and next will wave his handkerchief above the blind in like manner, then pull it back and rise as he shoulders his repeating shotgun.

In addition to the pictures and written accounts of retrievers thus far cited, there are several mid-nineteenth-century paintings of nondescript dogs being employed for duck hunting. A famous one by William Ranney, painted in 1849 and entitled *Duck Hunters on the Hoboken Marshes,* depicts two men in a pirogue with a dog bearing a strong resemblance to the modern Brittany spaniel. One of the men is thought to be the artist himself and the other a friend who sometimes hunted with him.

*William Ranney's 1849 oil, "Duck Hunters on the Hoboken Marshes," reveals changes in Hudson River as well as in retrieving dogs.*

The records allege that the first Brittany spaniels to reach the United States were imported by Louis Thebaud in 1912. But Brits had been used for centuries in Europe. Just as some retrievers perform well in the uplands, this spaniel serves nicely as a fetch dog in the water. Ranney may have portrayed a mongrel—mongrels rather than pure-bred retrievers must have ushered in the Atlantic Flyway's golden age—but the painter may just as well have been acquainted with the Brit.

Another aspect of his painting merits comment in connection with the golden age. It was one of the earliest genre oil paintings to show hunting as a sport in America, and without any need for further documentation in this book it reveals in an appalling manner what caused the demise of the golden age, what threatens the very survival of water birds on this flyway. The scene is set on the marsh waters off West Hoboken. All is serene, aglow with early-morning light, visually silent and expansive. Only one human habitation is in view, and that one only to an extremely perceptive observer who knows the region intimately. It is the barely suggested outline of St. Michael's Monastery on distant bluffs, hardly discernible. To the uninformed eye, it would be a natural irregularity of the Hudsonian heights. There is nothing else but water, vegetation, birds, two isolated men, and their dog.

Today, viewing the same panorama from the same point, one would see little water, few birds, hardly any live vegetation, and the very antithesis of isolation. The site of the monastery is invisible from there, the bluffs hidden by black factories, smokestacks, tenements, ware-houses, grimy office buildings, the bleak hangars that accompany piers and drydocks. And if these visual obstacles were leveled, the smog would intervene. The palisades of rock and greenery have been breached. What water remains is coated with debris and a light-refracting film of oil, a lethal curtain of motley over equally lethal sewage and organic mercury and pesticides whose half-life may be longer than the history of the village of Hoboken.

In these old pictures of an entirely bucolic flyway, the historian notes other less depressing clues to the past of American hunting. Some are clues of omission. Decoys are evident only in a few engravings and lithographs, not in the highly polished oil paintings. The meat-gathering settlers had assimilated the Indian's craft of the decoy and transferred it to the realm of shooting birds for market. But prior to the widespread acceptance of Jacksonian Democracy and the admission of the bourgeoisie to art via genre painting, artists did not depict such lowly pursuits as meat-gathering. Even when they began to portray hunting, they revealed it as noble sport and not the activity of meat procurers.

There is some evidence that decoys were known in Europe for several centuries, but their use was the exception rather than the rule. There was no perceived need of wooden blocks because live decoys—domesticated Judas birds, usually tethered but sometimes merely released, inevitably to return for feeding—were more effective lures than artificial decoys could ever be. Live decoys were kept in use in America by sportsmen as well as market gunners until 1935, the year when baiting and the use of tethered birds were both, at long last, banned.

But market shooters had begun to improve on the reed decoys and the stuffed-bird decoys of the red man at least as early as the first quarter of the nineteenth century. When Joel Barber was amassing his famous decoy collection, he came upon a snipe stool made in 1800, and he found a memoir published by one J. Cypress, Jr., in 1842, recalling the employment of duck decoys by the author's great grandfather at "the Fire Islands."

A live bird required care and feeding. It was particularly valuable as a lure since it combined the ultimate in realism with genuine, natural movement and voice, but some commercial shooters held that only one or a few live birds provided all the movement and voice needed. The attraction of these decoy birds could be tremendously enhanced by the addition of numbers, artificial numbers, floating and shore-perching statuettes. The art of decoy carving and painting was thus born among the relatively Philistine market shooters, unheralded and unappreciated by those who hunted only for sport and by those who depicted the sport.

Alexander Wilson's *American Ornithology*, published shortly before his death in 1813, contains possibly the earliest description of mallard decoys:

Wooden figures, cut and painted so as to represent ducks, and sunk, by pieces of lead nailed on their bottoms, so as to float at the usual depth on the surface, are anchored in a favorable position for being raked from concealment of brush, etc., on the shore. . . . Sometimes eight or ten of these painted wooden ducks are fixed on a wooden frame in various swimming postures, and secured to the bow of the gunner's skiff, projecting before it in such a manner that the weight of the frame sinks the figures to their proper depth.

Soon there began to appear occasional references to the use of decoys for sport. Among the earliest was a description in *The American*

*Shooter's Manual,* published in Philadelphia in 1827. Disapproval can be detected in the tone of the author, who preferred to remain anonymous: "Hogsheads are sometimes sunk into the mud on the flats over which the birds fly, in which the shooter and his dog are concealed. Artificial ducks and stool ducks [live ones] are also employed to decoy these poor birds, and no trick, or stratagem that human ingenuity can devise for their destruction, is left untried."

Another book published in Philadelphia three years later stated that, on Chesapeake Bay, decoys were little known and had not proved very successful—one surmises those decoys must have been crude, indeed—but by mid-century, references to decoys occurred, as Barber noted, "principally in connection with Canvas-back shooting as practiced on Chesapeake Bay." Printed reminiscences show that by 1853 the combination of sinkbox and decoy rig was sometimes a ploy of sportsmen as well as market shooters. In *The American Sportsman,* by Elisha J. Lewis, published in 1855, one reads about "this system" being used forty years previously on Chesapeake Bay, where it had been "introduced . . . by some of the experienced wild-fowl shooters from the vicinity of New York." These men, the author made clear, were gunning for the market and were "known to fill a small vessel with ducks in two or three days, which they immediately dispatch for the markets of New York, Baltimore or Philadelphia."

Henry William Herbert, writing under his famous *nom de plume* Frank Forester, mentioned decoys, blinds, and tolling dogs in his 1856 *Complete Manual for Young Sportsmen.* The Chesapeake method of duck shooting, he said, was to wait "behind screens erected for that purpose on the points and islands which they must necessarily pass and shoot them on the wing. Another method much employed in this paradise of duck shooters, is to toll the ducks, as it is called, while they are feeding along the shore, quite out of range, by means of a dog trained to gambol along the bank."

He also mentioned (and deplored) both punt gunning and the Long Island use of sinkboxes with a "whole fleet of decoys of all kinds and sizes." Forester declared that he was not alone in his disapproval of the sinkbox surrounded by a screen of decoys. "On Jersey waters," he wrote, "Squam Beach and Barnegat, and other places of equal resort of wild-fowl, prohibition of this destructive machine is on the whole enforced by the natives, a half-piratical race, half-fowlers, half-fishermen, and more than half-wreckers."

Whereas both sinkboxes and the cannon-sized punt guns met with opposition in a few areas where the market gunners feared that such contrivances would permanently frighten away ducks, the use of decoys was universally approved and imitated.

Both stick-up decoys (most often simulations of shorebirds) and floating blocks were occasionally made of cork or balsa, but the standard materials were white pine and cedar—cheap woods, generally available, easily carved, and durable. Carving and painting differed from place to place, and regional styles, or "schools," developed. Some have been so distinctive that a knowledgeable collector can frequently tell at a glance where an antique was made. For example, two-piece hollow decoys, cleverly engineered for light weight and proper flotation, were popular around Stratford, Connecticut, in the 1870s. Meticulously painted black-duck and broadbill decoys were a specialty of the Stratford School, and they were characterized by low, handsomely carved heads, equally well-shaped tails, and overhanging breasts designed to ride over slush ice on the marshes at the mouth of the Housatonic River.

On the Jersey coast, where hollow "dug-out" decoys were popular, the blocks were smaller than the Stratford type but with oversized heads. The small bodies facilitated the loading of some forty decoys on the aft deck of a Barnegat sneakbox. Guides and market gunners who favored these little boats customarily used a rig of about two dozen scaup and half a dozen each of blacks, brants, and Canadas. Though small, the decoys rode high on the water, and their big heads could be spotted by birds at considerable distances. The large heads also were durable, easy-to-grasp handles for tossing decoys about.

Off Nova Scotia, fishermen used "duck tubs"—sinkboxes surrounded by large spreads of coot and eider decoys. Since the birds were shot at almost all times of year, a great number of these decoys were painted not in the traditional black and white eider pattern but in drab browns and grays to represent postnuptial, or eclipse, plumage, and some were brown with white breasts to represent the autumnal moult. Unlike a black and white decoy, which copied the drake's winter plumage, the brown decoys mimicked both ducks and drakes at appropriate seasons.

On the coast of upper New England, the men who went out in cooting dories generally used flat-bottomed sea-duck blocks with heads mortised into the bodies. These decoys, which could withstand rough handling and heavy weather, were usually supplied with full, wide breasts for buoyancy. Their painting was blocky and simple, since subtlety is not a requisite for tantalizing scoter, eiders, or oldsquaws. Here and in some more southerly areas, conventional blocks were supplemented by shadow decoys—flat profiles mounted on boards. Sometimes the floating board was given the contour of a duck or goose as seen from above.

The oversized "slat goose," with a body resembling a barrel or overturned boat, probably originated on Cape Cod, where monumental floaters were used together with live decoys to attract birds passing at great distances. In Massachusetts and later in Virginia (and undoubtedly elsewhere) some of the slat geese achieved an extra degree of realism by means of canvas-covered bodies.

There were innumerable variations—merganser blocks with head-crests of horsehair; grass- or cork-stuffed bag decoys; pine-knot heads for brant blocks; floating and stick-up wooden gulls used on Long Island as confidence decoys with rigs of brant; loon blocks used not only as confidence decoys but for securing loon stew.

Around the Chesapeake Bay, probably the most popular wood decoys were solid white-pine canvasbacks and redheads. Painting was simple but not usually crude, and even the attachments were strongly characteristic of the region: a chunk of lead or iron ballast well aft on the bottom, and a galvanized iron ring stapled to the base for the attachment of the anchor line. These decoys were of standard size, but from Virginia's Back Bay down into Currituck Sound, giant canvasback and redhead decoys were favored; these, too, were said to have been introduced by market shooters.

As the century waned, the demand for decoys outstripped the ability of hand-craftsmen to turn them out. Most of the carvers were market gunners and guides who made decoys during the off-season and in their spare time. Some of these men, like the celebrated Elmer Crowell of East Harwich, Massachusetts, were clever enough to start turning out decoys in several grades—relatively crude, quickly finished blocks for volume sales (and often for their own use), better-made decoys only for those sportsmen willing to pay a higher price. Ultimately this practice developed into the uniquely American art of carving and painting decorative wooden bird statuettes for display rather than as "working decoys." Some of these figurines command high prices from devoted collectors, and a good decorative decoy is often so lifelike as to elicit comments about avian rivals of Pygmalion's Galatea.

Crowell carved many decorative birds and miniatures, but in his early years he was primarily interested in working decoys. Toward the end of his market-gunning days he produced some under contract to Iver Johnson, a company which then sold decoys as well as guns. There were also factories which had begun to specialize in machine-made decoys shortly after the Civil War. Among the well-known producers were Dodge and Mason, both located in Detroit, and Stevens in Weedsport, New York. Their lathe-turned, hand-finished decoys could not equal the quality of the completely hand-crafted blocks made by some of the outstanding artists, but the old factory products did have a character lacking in the most lifelike plastic decoys of the modern era. Before turning to plastic, manufacturers tried papier mâché, cork, rubber, and other materials; some of their efforts were adequate but none matched the carved and painted variety.

Just as a school or area of origin can be detected in the carving and painting of a fine decoy, the work of an individual artist is often evident. Browsing through *The Classic Decoys Series* and *Classic Shorebird Decoys,* two portfolios of prints by the fine sporting artist Milton C. Weiler, the student can recognize the sure touch of Albert Laing in the construction of an 1870 white-winged scoter decoy. Laing was the founder of the Stratford School and may have been the first to design the high-breasted decoys which were so effective on Connecticut waters. He was also the first to use copper nails in fastening together the halves of hollow white-pine bodies. This eliminated the splitting and staining of wood caused by the corrosion of iron nails in salt water. His work influenced the two great Stratford makers Ben Holmes and Shang Wheeler. Holmes entered a dozen broadbill decoys in competition at the Centennial Exposition of 1876 in Philadelphia, and he won all the honors. One of Weiler's prints depicts an 1890 broadbill drake by Holmes, and its smooth mastery shows how a single carver was able to dominate such an exhibition.

A blend of realism and graceful modeling, evident in another plate, a 1920 black duck by Shang Wheeler, reveals why the last of the great Stratford triumvirate is considered by some connoisseurs to be the master sculptor among decoy craftsmen.

One of the earliest of the great carvers was Nathan Cobb, a New Englander who was shipwrecked on the Eastern Shore of Virginia in 1837 or 1838 and decided to settle where destiny had tossed him. He introduced to that region the hollow-body technique of the pioneering Long Island and New Jersey makers and added to it highly individual flourishes of realism. The wooden geese, brant and ducks emanating from Cobb's Island were often sculpted in swimming position, with forward-thrusting heads of nearly indestructible holly root.

Among Cobb's contemporaries was Captain John T. Corwin, of Bellport, Long Island, a market shooter who could impart genuine personality to a decoy. Ed Zern, in text accompanying the Weiler prints,

White-Winged Scoter - 1870
Albert Laing..... Stratford, Conn.

*Prints from Milton C. Weiler's superb portfolios of classic wildfowl and shorebird decoys evoke trends, innovations, and artistic triumphs of bygone years.*

Whistler Drake - 1870
Capt. John J. Corwin.. Bellport, L. I.

Black Duck - 1920
"Shang" Wheeler..... Stratford, Conn.

"TINNIES"

HARRY V. SHOURDS
Tuckerton, New Jersey

remarked of an 1870 Corwin whistler that it "swam with the same perky élan a live goldeneye would show."

Another fine artist of the 1870s was Henry Grant, in whose work can be seen the full flowering of the New Jersey style. In describing a Canada by Grant, Zern was moved to declare that "there is, in the best of the Barnegat Bay school's goose decoys, a classical cleanness of silhouette and the same simple but monumental spatiality that distinguishes most of the sculpture of Noguchi and some of Henry Moore's." A less sweeping but nonetheless massive simplicity can be seen in fine turn-of-the-century decoys such as those made by Ira Hudson, a Chincoteague boatbuilder. A bufflehead drake by Hudson is felicitously described as evincing the same "taut alertness . . . chunky seaworthiness" that characterized the boats he built.

The tradition of perfection and originality was carried into the twentieth century by men like Elmer Crowell and Joseph Lincoln, both of Massachusetts. A 1900 widgeon by Crowell and a 1920 wood duck by Lincoln seem far too beautiful to be subjected to the rough handling of the hunt.

Few carvers of this rank are left, though superlative exhibition decoys are produced by such men as Bill Birk of Bridgeport, Connecticut, Arnold Melbye of South Yarmouth, Massachusetts, Dan Brown of Queenstown, Maryland, and Cigar Daisey of Chincoteague, Virginia. Even the most famous of contemporary makers, the brothers Lem and Steve Ward of Crisfield, Maryland, long ago began to specialize in ornamental birds. To carve and paint gunning decoys today, in competition with the plastics manufacturers, would be absurdly unprofitable. As a hallmark of handmade decoys, appearance has replaced hunting quality and basswood has replaced pine and cedar.

Today's most successful carvers are known exclusively for their ornamental birds, even though most of these men—Davison Hawthorne, Oliver Lawson, Kennard Massey, Ronald Rue, Wendell Gilley—live in the best gunning spots on the Atlantic Flyway. In that list, all except Gilley (whose home is at Southwest Harbor, Maine) are from the same area as the Wards.

Not very long ago Dan Brown emulated the prize-winning feat of Ben Holmes by entering eight birds in the National Decoy Competition and capturing top honors in assorted categories with all eight. In a manner of speaking he has also revived a quite different tradition by his consummate skill at carving shorebirds. Some of the old-time artists who gained renown with their duck and goose decoys also produced exquisite likenesses of shorebirds in the days when the appropriate species were among the prime quarry of both sportsmen and market gunners. Among those who come to mind are Cobb, Hudson, and Crowell.

Most shorebird decoys were of the stick-up type, and the best of them often exhibited even more individuality than the floating blocks. As with duck and goose decoys, there were regional characteristics; the Nantucket style at any given period might be quite different from that of Great South Bay or Chincoteague.

Folding tin shorebirds, patented by H. Strater and W. Schier of Boston in 1874, were stamped and marketed solely for the delectation of sportsmen from that year until 1918, when federal law terminated the havoc then being wreaked on shorebirds. Tin decoys could be folded and stacked compactly, and they were well received by sportsmen, but professional gunners felt that four dollars a dozen was too much to pay for gadgetry, especially gadgetry that quickly rusted, and so they remained loyal to wood.

One of the finest carvers who produced both wildfowl and shorebirds was Harry V. Shourds of Tuckerton, New Jersey. Shourds was born in 1861 and he died in 1920. During his prime he exemplified the originality and perfectionism of the Atlantic Flyway sculptors. William J. Mackey, Jr., author of *American Bird Decoys* and the text for Weiler's *Classic Shorebird Decoys,* described the qualities of a Shourds masterpiece in the latter work:

The chaste, practical simplicity . . . cannot be faulted, and so expert was his selection of wood that an age crack is almost unknown in any of his work. White juniper was always used for the bodies, and white pine was whittled into the inimitable heads of his ducks, brant and geese. One minor deviation should be noted: one rig of Canada geese that was made of mahogany. All of his snipe are of one piece and carved from cedar. The bills are splined white oak.

Such carving is extremely time-consuming, particularly for a man whose sculptures are, in Mackey's phrase, "dainty to the point of elegance," yet Shourds stole sufficient hours from his nominal trade of painting houses to make more decoys than any other man, and he did so entirely with hand tools. Such dedication to one's true calling is understood by a few artists, perhaps some clergymen, and most duck hunters.

Many of the gunners who relied on the old wooden decoys enhanced the attraction of a rig by a judicious scattering of bait. A great many kinds of bait have been used, including tomato seeds, weed seeds, and even crushed stone in areas where birds might be lured by the prospect of grit. Although it has been suggested by more than one authority that milo is probably the food most desired by ducks, corn always was and still is the chief offering.

Technically, federal law put an end to baiting in 1935, but the offense remains so regrettably widespread that airplanes and helicopters are used by wardens to survey areas where greedy or tradition-bound shooters are believed to be incorrigible. Fortunately scattered corn (and thick congregations of fowl in unlikely places) can be seen from high in the air. Not so fortunately, baiters know that the most effective technique is to keep feed plentiful in an area for an extended period before and during the hunting season, and to refrain from shooting too often in the area. This reduces the chance of apprehending a culprit in the act. Furthermore, there is no law against leaving standing corn in a field next to a gunning site. It might almost be claimed that the entire mid-Atlantic corn belt is one vast baited gunnery. But there are two comforting factors in this situation. First, the corn (except when purposely used for baiting) is of incalculable benefit to wildfowl where former natural habitat has disappeared. Second, sportsmen have become so sharply aware of the need for conservation that baiting is at last succumbing to the disapproval of the gunners themselves.

For many years, baiting lured ducks not only to the guns but to traps set by the purveyors of game meat. Many kinds of traps were used, but they were most commonly rigged in shallow water with a frame of stakes or poles and a covering of net, cord, or wire. One well-baited and well-disguised trap caught sixty-seven ducks on a single full-moon night while the marketer tended to other affairs. The federal ban on trapping in 1918 was one of the important elements in ending the slaughter of birds for the market. Although traps have been confiscated and trappers arrested as recently as the 1960s, the crime is now virtually extinct, a fate that might have befallen several gregarious species of birds if the fury of sportsmen and other conservationists had not intervened in time.

Together with baiting, live decoying was federally outlawed in 1935, long after a number of states had legislated against these practices, and the final retirement of the tethered goose had a strong effect on hunting techniques. One of the men who perfected the most devastating techniques with bait and decoys was the legendary decoy maker Elmer Crowell. Most of his gunning was done during the late nineteenth and early twentieth centuries—the zenith of the golden age, when market shooting for a few species had already ceased to be profitable but guiding wealthy sportsmen had become a lucrative business.

His home at East Harwich, on Cape Cod, was at the hub of the region made famous by the ''Massachusetts goose stands.'' These were extremely long blinds erected on the shores of bays and ponds, and they were of unusual construction. A typical stand consisted of a shoulder-high board fence, camouflaged on the side facing the water. On the side where the shooters stood were nails or pegs for hanging birds and gear, and there was usually at least one long shelf for additional gear and ammunition. At one end was a camouflaged hut where wooden decoys and other equipment could be stored, and where shooters could find occasional brief respite from wind, cold, rain, or snow. At some distance to the rear, preferably on a hill or cliff, were pens for live decoys, of which as many as a hundred were used by some guides toward the end of the era. Often these birds were so well trained that there was no need to tether them, and they were flown from the cliff to mix with passing flights and coax the wild geese in over their wooden and live tethered counterparts. The decoy birds abandoned their luckless dupes at the last moment, and a winged Judas generally enjoyed a long, well-fed life of treachery.

There is no doubt that the first stand shooters were gunning for marketable meat, but their singular edifices eventually drew the attention of visiting sportsmen. The market gunners of eastern Massachusetts began to supplement their incomes by serving as guides; some of them built lodges within reasonable distance from the stand sites, and others erected stands near existing lodges, where the professionals were hired as caretakers or "gamekeepers." Their shooting stands became an important sporting attraction in Massachusetts.

In 1947, when he was nearly eighty-five years old, Elmer Crowell contributed a brief memoir to a fine volume edited by Eugene V. Connett entitled *Duck Shooting Along the Atlantic Tidewater*. In this reminiscence, he described his use of live decoys and bait as well as a stand which he erected in 1876 and used for thirty-two years of market shooting. When he was fourteen, his father bought a tract of land on Pleasant Lake at East Harwich. It had a sand beach that was ideal for tethering decoys, but he owned only six of them that first season. Using them in conjunction with nine block decoys which he set out thirty yards from shore, he killed about a hundred ducks and then gave some thought to improvements for the next season. When the time came, he had twenty-eight live decoys.

I pinned six on the beach, and kept the twenty-two for flying from the blind. The blind was a board fence about thirty yards long, and when well brushed with small pine and oak limbs looked very much like the background. I had very little trouble in getting the birds near enough to shoot. That season I killed one hundred and eight Blackducks and many small ducks.

During the next few years, he also killed some shorebirds for market but noted that they were already declining. His father had told him about a day in his own youth when he had seen two men fill a bushel basket with golden and blackbellied plover in one morning. Those birds were no longer plentiful and he concentrated primarily on greater and lesser yellowlegs, plus a few Eskimo curlew which were locally called doughbirds because they were so fat. But he continued to direct his greatest efforts at ducks and geese.

*Henry Watson's illustrations for Grover Cleveland's "Fishing and Shooting Sketches" (above and opposite) showed use of live decoys.*

The next season I had forty live decoys and tried out a new way of handling my ducks. I sank a pole with a ring in the top and ran a line through it back to the blind, which made an endless line. I tied four live ducks to it and pulled them off in the pond about thirty-five yards. It was about two weeks before they were broken in so I could pull them out without their making a flutter. After that they were all right.

Then I began to fly them from the blind. It was some job to get them back to the blind at first, but in a few weeks they were broken in fine. They would fly out to the runner and light with the four ducks made fast on the endless line and they would follow them to the beach. So when a flock of wild ducks came along we filled the air with decoys and the wild ducks would light right with them. After that, it was easy to pull them to the beach and within shot. That was the first time it had been done with any success.

*A Duck Hunting Trip.*

Although live decoys had been used for many years in America and for centuries in Europe, Crowell became an innovator of release and control methods. A couple of years after he had increased his tame flock to forty birds, a sportsman named Charles Hardy and several friends built a lodge called Three Bear Camp on Pico Point. Crowell was entrusted with the care and handling of the decoys. He used about fifty live geese, of which twenty were tethered on a large beach that was built out into the lake for that purpose. Pens were constructed on a hill in back of the blind, and electric wires were run from them to four boxes of control buttons in the blind. When wild geese were sighted the shooters pressed the buttons, a trapdoor fell, and out came the flyers, rising over the lake and sometimes bringing the unsuspecting quarry right back over the beach. In addition to his live birds and blocks, he employed huge slat geese to attract the attention of distant flights. The guide recalled:

It was a success, and we had good shooting for a number of years. At that time we began to bait the small ponds with corn, and two years later we baited in front of our blinds at the lake. It stopped the Blackducks from going south, so the ponds were full of ducks; the shooting was great. But we could not sell them in the markets, as the law cut it out. Soon the law cut out the live decoys, and that was the end of good shooting here.

Of course, Crowell's definition of good shooting would seem more like slaughter to a modern sportsman. By 1913, when the federal government established its right to regulate hunting seasons for birds migrating across state borders, Crowell was devoting less of his time to shooting for the market and more of it to guiding and making decoys. In 1918 when the Migratory Bird Treaty was signed, he quit shooting. After his son Kleon returned from service in World War I, Crowell built a new blind on Bushey Beach Pond; father and son gunned there

## ALLEN'S DECOY DUCK FRAME.

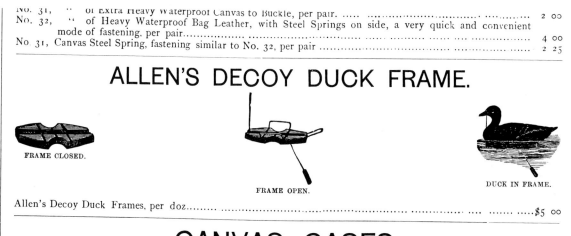

FRAME CLOSED.

FRAME OPEN.

DUCK IN FRAME.

## CANVAS CASES.

*Decoy gadgetry has been multitudinous and strange;
this example is from 1884 catalog of sporting-goods store.*

for several years. Then Massachusetts prohibited baiting, and the senior partner sold the blind. Both Elmer and Kleon Crowell became famous for their decoys, but the old man seldom did any gunning after that.

He began to specialize in carving and painting decorative pieces, which he continued to produce until rheumatism ended his career when he was past eighty. Long before he died in 1952 at the age of ninety, he had become a Cape Cod legend, dispensing wildfowling lore and decoy-making advice to all who sought his counsel. One cannot condemn such a man for remaining at heart an unregenerate market hunter who could write that "my mind wanders back to the good old days when there was no law on birds." Having seen the immense flights sweep down from the north year after year, he was unable to grasp the concept of restrictions on what seemed to be an inexhaustible supply of game. He would have been horrified at the idea of endangering any species and yet he could not adjust to the new age of self-imposed restraints.

One of these restraints, the cessation of live decoying, spurred the popularity of a formerly unimportant hunting tool, the wildfowl call. The Indians had relied on vocal mimicry, without any mechanical aid, to call birds into range. Today, many Indian guides in the North can still produce an astonishing imitation of a goose, as can a few white sportsmen who have been practicing since boyhood, and there are gunners all along the flyway who can elicit answering calls when they purr to a flock of broadbills. But until the demise of the flying or tethered decoy birds, most of the calling was done by those birds. Sometimes a tame goose or duck became temporarily taciturn, but Crowell and most of his colleagues knew how to coax a bird into conversation. Like the Indians, they generally used no mechanical aid. They simply used their voices to mimic a feeding or gathering call. As a rule, the live decoys answered, and the chatter, once started, tended to spread and continue.

There were, of course, some hunters who experimented with the whittling of mechanical calling devices. No one knows who first fashioned a reed or cane instrument that vibrated at the proper pitch when blown like a whistle, but the Cajun hunters of the South, where the Atlantic and Mississippi flyways blend, long ago became famous for their small, slender cane calls. And there is no doubt that other gunners experimented with crude wooden calls in which a reed or stiff grass blade vibrated to produce an anserine sound.

Mechanical calls have been used on the Atlantic Flyway for at least two centuries to lure the wild turkey into range. In this type of hunting, too, the vocal cords were (and sometimes still are) employed without any artificial aid. But the settlers quickly discovered that a turkey's wing bone could be held to the mouth to add the desired vibration, and even without benefit of any authenticated records it is difficult to imagine that duck hunters did not attempt the same kind of experiment. Box calls also appeared early among turkey hunters. The best of these were made of basswood or yellow willow, though cedar could be used where those woods were scarce. One of the two most popular types utilized a chalked piece of slate which was rubbed across the top edge of the box. The other had an attached cover, beveled on the underside and chalked at the edges, which engaged lips at the top of the box and could be rasped along to emit an excellent impression of a turkey. Such calls are still being made and used.

There is no doubt that calls of this type were tried by duck hunters, too. Unfortunately, no variation on the box call produces a very real-

*Modern stick-up decoys are extremely effective in goose fields.*

*Today's mass-produced blocks are less artistic than hand-carved decoys but are durable and efficient.*

istic duck or goose sound, so wooden mouth calls were usually whittled by gunners who wished to overcome the difficulty of imitating the birds by voice alone. One finds occasional references to homemade calls, and there are old-timers who remember the tedious trial and error of shaping and adjusting the reed to get the right tone—as well as the frustration when a loose, wet, or old call unexpectedly gave forth sour notes.

The late nineteenth century saw the rise of the successful business-man as an American folk hero. He had the leisure to go hunting, and had shed much of the Puritan guilt associated with diversion. The new sportsman was not the type to whittle reeds and wooden cylinders, but he enjoyed store-bought gadgetry. The business of supplying equipment and accessories quickly became a lucrative enterprise. Inexpensive mouth-operated duck calls—and shorebird calls, too, for that matter— have been sold in American sporting-goods stores for at least three-quarters of a century. In his *Fishing and Shooting Sketches,* published in 1906, Grover Cleveland emphasized calling as the key to successful shorebird shooting. The former President advised:

The initiated . . . know that the best caller will get the most birds. The notes of shore-birds, though quite dissimilar, are in most cases easily imitated after a little practice, and a simply constructed contrivance which can be purchased at almost any sporting goods store will answer for all the game if properly used. The birds are usually heard before they are seen, and if their notes are answered naturally and not too vehemently or too often, they will soon be seen within shooting range, whether they are Black-Breasted Plover, Chicken Plover, Yellow Legs, Piping Plover, Curlew, Sanderlings or Grass Birds. Of course, no decent hunter allows them to alight before he shoots.

Cleveland's testimonial notwithstanding, there was an evident lack of confidence in the durability and reliability of commercial calls. Too often they lacked a realistic tone to begin with, but by the 1920s, when various states were outlawing live decoys, several firms were success-fully marketing calls of improved quality. References to such devices became more frequent in writings of the period, and after the live-decoy ban became federal in 1935, factory calls proliferated.

Before World War II, the hard-rubber duck call made its appearance. Even at its modest price of fifty cents, it was less than completely satisfactory. An acceptable crow call could be made of hard rubber, but no manufacturer seemed able to produce a rubber call with a really fine duck or goose tone.

Several companies returned to the use of more traditional materials, mass-producing wooden cylinders which enclosed reeds. Such calls have remained most popular. A more recent development has been a call operated by manipulating rather than blowing it. This device employs a pleated cylindrical bellows—a sort of miniature rubber concertina with a short gripping section at each end. There are small models which emit quacks and chattering calls when compressed and expanded, shaken or twisted, larger ones which imitate the more magisterial bugling of a honker. As every wildfowler has been forced to admit at one time or another, silence is preferable to poor calling. The bellows instrument provides good mimicry and has the advantage of being easier to master than the breath-operated call, but it is also less versatile. In the hands of most gunners, its utterances lack the subtle nuances, the finesse, of a mouth call operated by an expert.

And so the traditional wooden mouth call has retained its ascendancy, hardly changed except for the replacement of a natural reed with a synthetic one. Many a sportsman would not dream of spoiling his chances by blowing a call he has owned for years and never mastered, but neither would he dream of entering a blind without that call dangling from a neck thong, for it has become an element of his hunting garb and marks him as a wildfowler.

*After many decades of experimentation,*
*mouth-operated wooden calls remain most popular.*

# Chapter Nine

# WAYS AND MEANS OF THE WATERMEN

In 1621, Edward Winslow of Plymouth wrote home to advise fellow Englishmen about equipment needed by those who would settle the New World. Among his less puritanical countrymen, duck shooting had already become a minor sport as well as a method of food gathering, and gunners on the English coasts positioned themselves in blinds—usually called stands or huts—along flyways or near feeding areas. Regarding suitable firearms for the colonies, Winslow advised, "Let your piece be long in the barrel and fear not the weight of it, for most of our shooting is from stands."

It is hardly surprising that the early ducking guns have come to be known as long fowlers, for they averaged six to eight feet in length and weighed fourteen to eighteen pounds. Gunmakers of the seventeenth century were aware that gunpowder could exert its propellant force only while the shot charge was confined in the barrel. The slow ignition of their flintlocks and the burning characteristics of their crude black powder supported the axiom that "the longer the range, the longer the barrel." The powder-burning efficiency thus attained might be negated by the friction to which a shot charge was subjected in an extremely long barrel, but no one recognized this countering effect.

The first fowling pieces to reach the Atlantic Flyway were therefore spectacular in size and weight, and it is with great respect for the gun-handling prowess of the Pilgrims that one reads old accounts of how they astonished the Indians in the 1630s by killing crows in flight. The common depiction of our forefathers blasting away at tightly massed "sitting" ducks on the water is not quite accurate, for our forefathers were practical men. Anyone who has had to deliver the *coup de grace* to a downed but live duck knows that a bird on the water presents a small target, low to the surface. And because of the angle of fire, most of the shot pattern uselessly spatters the water. It is easier to kill a duck in the air.

When the object of the hunt was to procure meat, early gunners tried to kill more than one bird with a single shot. Market shooters often brought down half a dozen with a single pull of the trigger, and there are authenticated records of punt guns killing seventy-six sprig at a time and on one occasion eighty-six canvasbacks—in fact, as many as a hundred ducks—though a good shot from such a gun averaged about thirty ducks or eight to ten geese. But those birds were not shot on the water, nor could they have been. The closest ducks would have shielded those farther away. When the ducks were sitting, the gunner shouted in order to flush them, and he fired when they had risen a couple of feet above the water. As for seventeenth-century gunners, those who could afford the luxury of sport frequently preferred true wingshooting, or "shooting flying" as the English called it.

Some of their long fowlers had bores as small as .53 caliber (not quite 28-gauge) but larger ones were more common and they ranged all the way up to .80 caliber (a bit larger than 10-gauge). A big single-barreled fowling piece of this kind could be loaded with a single ball for deer or even bear in the new wilderness, and it could be loaded with "hayl

Colonists brought long fowlers like that used by John Thompson (above) and another early settler, John Forbes (below).

shot" for birds. Thickly fortified at the breech, it could fire a heavy charge without danger of bursting and it was used not only to procure meat but to fend off Indian attacks.

Among the newly arrived colonists at Plymouth in 1622 or 1623 was John Thompson, whose .84-caliber long fowler still survives. Tradition has it that Thompson did not consider himself a skillful shooter, though he was proud of his ability to lift and aim his twenty-and-a-half pound fowling piece without any support beneath its six-foot barrel. He served in the Plymouth militia, carrying his long fowler and "brass pistol" during several clashes with French settlers to the north and then in 1673 on a bloodless bluffing expedition against the Dutch to the south.

Two years later, during the short but furious Indian uprising known as King Philip's War, he had charge of sixteen men defending the little Plymouth garrison when Philip's Wampanoag warriors attacked. He handed the fowler, loaded with a ball, to a noted marksman named Isaac Howland and instructed him to shoot a brave poised on a distant boulder. The Indian, apparently believing himself to be safely out of range, was standing in the open taunting the defenders when Howland shot him down. After a further display of long-fowler marksmanship by the Pilgrims, the war party retired.

It is safe to say that such duck guns were numbered among the most essential equipment of the settlers. Supplies brought to the Massachusetts Bay Colony in 1626 included half a dozen long fowlers with six-and-a-half-foot "musket bores" (about .75 caliber) and four more with five-and-a-half-foot "bastard musket bores" (.69 caliber). Most such arms at first employed dog locks, simple and relatively crude flint-operated mechanisms named for a small "dog," or catch, that afforded a half-cock safety position. These were soon supplemented by English locks of the sort Thompson owned, with a laterally moving sear and internal half-cock notch, and the true flintlock gained prominence shortly after mid-century.

The guns soon began to exhibit distinctive Atlantic Flyway contours unlike the style of the first, purely English fowlers. From the Connecticut coast up to the southern shore of Cape Cod, the buttstock developed a graceful drop similar to that seen later on the "Kentucky" rifles perfected by the gunsmiths of Pennsylvania. Normally these smoothbores still employed English parts, but the Dutch tradition heavily influenced the gunsmiths along the Hudson-Mohawk River flight lanes

*Typical long fowler and punt gun, seen in 1822 English caricature, evolved in Britain but soon became established on America's East Coast.*

# SHARERS
# OF
# THE
# WETLANDS

Black ducks are Atlantic Flyway favorites, as are three superb retrieving dogs seen here—proud, gentle-mouthed golden, eager Chesapeake leaping into water, black Lab returning with mallard.

Splashing wood ducks bring to mind Thoreau's lines
about "wild and noble sights" denied to those
"who sit in parlors." Hunter arranging goose blocks
may wonder what spectacles he will see on Eastern Shore
where gunners once floated giant wooden swans.

*Mallards alight near shore favored by waddling coots and statuesque yellowlegs. Alert sanderling and red-backed sandpiper on foamy beach are wooden, carved by Davison Hawthorne.*

*Gaping beak and cocked neck give egret argumentative air, and stride of great white heron seems purposeful, but wise observers are always suspicious of anthropomorphic interpretations.*

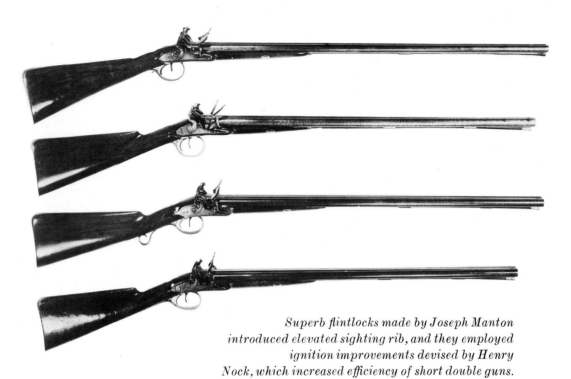

*Superb flintlocks made by Joseph Manton introduced elevated sighting rib, and they employed ignition improvements devised by Henry Nock, which increased efficiency of short double guns.*

and into New Jersey. These makers favored a fat-bellied stock quite Dutch in appearance except that it was often fashioned from curly maple and sometimes embellished with fine relief carving. Dutch mounts and barrels were frequently employed even as late as 1700, but within another half-century long fowlers had virtually disappeared except for a few that were treasured by early market shooters.

The cumbersome, muzzle-heavy guns were replaced by lighter, better balanced pieces—still heavy and unwieldy by modern standards but often very well-made and occasionally displaying exquisite decoration. Then, in 1787, the great English gunmaker Henry Nock designed a new breech which offered much faster, more efficient ignition than that of previous flintlocks. With the new breeching, shooters discovered that a barrel could be as short as thirty-two inches or even less without any sacrifice of range or power.

Double-barreled guns had been built before with forty-inch barrels—substantially shorter than long fowlers but too awkward and muzzle-heavy to gain wide acceptance. Now, however, the short, light double gun had arrived on the scene, and its establishment was accelerated by Joseph Manton's 1806 invention of the elevated sighting rib to increase long-range accuracy and by the Reverend Alexander John Forsyth's almost simultaneous invention (soon afterward perfected by others) of the percussion ignition system.

In 1812, Johannes Samuel Pauly, a Swiss inventor living in Paris, introduced a break-open shotgun which pivoted on an under-hinge and closed against a standing breech. Truly practical breechloaders required but a few more decades of refinement; they appeared, together with fully developed self-contained ammunition, in the 1860s. Within ten more years, such great British makers as Rigby, W. W. Greener, and Anson and Deeley were offering "hammerless" (or concealed-hammer) models with top release levers and snap-closing latches. Such arms found their way to the Atlantic Flyway and then westward throughout America almost as soon as they had appeared in Europe.

This was also the period when choke boring was discovered, or rediscovered. In 1875 an English publication, *The Field,* awarded a prize for the invention to W. R. Pape, of Newcastle, on the basis of an 1866 patent. Most American scholars prefer to credit the invention to Fred Kimble, an Illinois duck shooter who won renown as the greatest wildfowler who ever lived. (In the late 1870s, Kimble also developed a clay pigeon to replace the live pigeons and glass-ball targets then in use, and his choke-bored shotguns helped him to become the most successful competitive trapshooter of his day—never beaten in a match.)

Kimble himself never claimed credit as the inventor of choke. In a brief biographical sketch published in 1961, Charles B. Roth, who had known Kimble, wrote that the Illinois hunter had stumbled on the idea of muzzle constriction almost accidentally while tinkering with a single-barreled muzzle-loader that refused to throw a good shot pattern far enough to suit him. That was just after the Civil War. Many years later Kimble told Roth he probably would have abandoned his experimental tinkering if he "had been a better scholar . . . because years afterward I learned that as early as 1781 British gunmakers were experimenting with the same idea. They had given it up as a bad idea."

Kimble's discovery was the result of exasperation at the inability of a nine-bore muzzle-loader to hit mallards flying over forty-yard-high trees along the wooded river bottoms where he hunted. He first tried reaming out the muzzle to achieve a mild blunderbuss effect, and when that failed he bored the rest of the barrel to a slightly larger diameter, from the breech to within an inch of the muzzle. He overdid it and the first firing test was disheartening, but he decided nonetheless that "it made sense . . . if you would constrict the muzzle . . . you would get a smaller pattern."

After a bit more tinkering he discovered a degree of constriction that reduced pattern size at a forty-yard range with an ounce-and-a-half load from a five-foot scattering to a circular area of about thirty inches. Obviously, the Englishmen who had toyed with this concept eighty-five years earlier should not have given up so easily.

Kimble's gun soon won a reputation in Illinois, where every community boasted at least one "world's champion duck shooter," and he

In 1880 lithograph from painting by L. E. Earle, stylish sportsman proudly shows cartridge-firing double gun to less affluent farmer, who has not done badly with dilapidated single-barreled muzzle-loader.

found himself matched against a revered gunner named Reese Knapp.

Knapp, he recalled, used a gigantic single-barreled four-bore muzzle-loader. "The barrel was four feet long. And the gun weighed sixteen pounds. Knapp's usual load was two and a half ounces of shot. The only gun I had was my little 9-gauge and I used only one and a quarter ounces of shot. By the way, Number Threes were my favorite, soft lead shot, of course; we didn't have anything else."

At the end of the match, Knapp's huge fowling piece had bagged thirty-seven mallards; Kimble had brought down a hundred and twenty-eight.

Regardless of who invented choke, most historians cite W. W. Greener as the one who perfected it, allegedly in about 1874. Perhaps credit for its perfection does belong to England, but by then choke-bored guns were in use on the Atlantic and Mississippi flyways. Kimble had shared the secret of his ducking success with Joe Long, a shooting crony in Boston. And Long instructed Joseph Tonks, a famous Boston gunsmith, to build a fowling piece for him on the Kimble formula. Tonks produced two, one for Long and another—intended for a customer who died—sold to Kimble. After that, no more than word of mouth was needed to awaken a demand for choke-bored guns along the flyways.

Thus the "classic" duck gun—the choke-bored double-barreled breechloader—was a rather late arrival, having come to prominence in the last quarter of the nineteenth century. It is still regarded reverently, but perhaps more as a status symbol than for its inherent merits. It is more likely to be seen in a cabinet, polished and protected, than in a duck boat, flecked with corroding spray.

Though a fine over-and-under gun is slightly less costly to manufacture than a side-by-side double of comparable quality, and is increasingly popular, the over-and-under configuration has never usurped the place of the traditional side-by-side. However, doubles of both kinds have largely been replaced by repeaters, for reasons that can be ascertained from the very nature of American wildfowling.

Undeniably, there was a brief gilded era when clubs of wealthy sportsmen proliferated along the Atlantic Flyway, and since then more and more of the finest gunning sites have become private property. Those facts notwithstanding, what began as foraging for meat has never evolved into a sport exclusively, or even mostly, for the landed gentry. If duck shooting enjoys an aristocratic reputation, the aristocracy is primarily of skill, not money. Its trends have always been influenced strongly by the preferences of the most successful duck hunters, few of whom have been wealthy. They have been tough, practical working men, the watermen of the seaboard who have earned their livelihood by fishing, whaling, lobstering, crabbing, digging oysters and clams, piloting vessels, shipping, smuggling and bootlegging when times were hard, farming, milling, manning the factories, hunting shorebirds and upland birds and furred game and—preferably—ducks and geese.

For many of them, wildfowling was both vocation and avocation. Those who hunted for market supplemented their incomes by guiding both resident and visiting sportsmen. They designed and built the gunning boats and blinds, perfected the art of decoy making, devised the hunting methods.

It is in his capacity as a teacher that the guide has always been most highly valued by the amateur who hunts purely for pleasure. The waterman's choice of guns, therefore, has been regarded as a mandate by the majority of wildfowlers. A visiting hunter generally had an opportunity to watch his guide shoot, and the guide seldom missed. If a tight wedge of broadbills or teal flew close, or a thicket of sprig hurtled into the air, it was not uncommon for a guide to fell seven at one blow. He might, in fact, hesitate to expend a shot for so paltry a return as one little teal—a mere "breakfast duck"—but half a dozen could mean dinner for his family or a couple of dollars in his pocket. After watching a man like that in action, almost any sensible wildfowler would emulate his choice of guns as well as his shooting style.

So it was that the watermen of the Atlantic seaboard influenced the trends of American gun design. The handling qualities of the double gun—its good balance for aim and swing, its fast maneuverability, its convenient size—made it far superior to the old long fowler, as did its

second charge of shot. Its quick ascendancy was inevitable, but so was its demise. Some arms historians have speculated, and with good reason, that the American penchant for repeating shotguns originated among the guides and market shooters of the Atlantic Flyway.

First they tried the Winchester lever-action shotgun designed by the gunmaking genius John M. Browning and introduced in 1887. It was unwieldy but it was followed in 1893 by the Browning-designed Winchester slide-action which, in slightly refined form, became the Model 97. This exposed-hammer pump gun remained in production for many years, and a specimen is still occasionally seen bringing down ducks. Its owner would regard any thought of parting with it as blasphemy. At the turn of the century, Browning followed his triumph of the pump with the first successful autoloading design. His square-backed old Remington Model 11 semiautomatic, like his Winchester Model 97, still turns up in duck blinds.

These guns functioned reliably and handled almost as well as a good double, but their spectacular success was simply a matter of five or six fast shots rather than two. Before federal regulation limited cartridge capacity for wildfowling to three shells, the magazine tubes of repeaters were never plugged. In fact, there were many incorrigible market shooters who continued to rely on the unplugged magazine long after it became illegal, and at least one of these men, on Chesapeake Bay, fashioned a magazine extension which made the tube fully as long as the barrel and increased its capacity to eleven shells.

In the course of research for his book *The Outlaw Gunner,* published in 1971, Harry M. Walsh interviewed a number of survivors of the market-gunning era. "Atley Lankford," he wrote, "has gunned all his life on the famed Elliott Island marshes. He gunned commercially from 1900 to 1918. The limiting factor in his daily kill of two hundred birds was, simply, carrying them off the marsh. Atley averaged better than ten thousand a year with a lifetime average of nearly half a million ducks. He has probably killed more ducks than any man alive." This kind of shooting wore out a gun per year until, in about 1910, Lankford acquired a semiautomatic which he was able to use for eight winters by replacing the wooden forearm each year. Lankford has estimated that this gun alone killed over thirty-five thousand ducks.

Another old market shooter told Walsh that "before the automatic shotgun we had nothing to kill them with." Even the gigantic punt guns could not compare, and their importance has been exaggerated since relatively few were used and for a relatively short period. "Around the turn of the century," according to Walsh, "at the height of the big gun's popularity, there were probably less than one hundred of these guns in operation over the entire [Chesapeake] Bay."

The professuonal gunners of Maryland, Delaware, and Virginia—highly successful ones like Ralph Murphy of Charlestown and Sam Armstrong of Delaware City—used punt guns only occasionally and turned from doubles to repeaters as soon as the new models became available. On the Rappahannock River, where shore batteries of over-sized guns were prevalent, a market gunner named Eagle Creighton matched his new autoloader against a duck cannon mounted on the shore of Hooper Island. He killed twice as many ducks as the big gun in a single day.

One day in about 1917, when the ducking had been poor in North Carolina, Lloyd Doxey and several other guides brought their sinkboxes up from Currituck Sound to Craft's Landing in Back Bay, Virginia. Doxey felt that the local shooters were not using big enough stools, so he put out more than three hundred decoys and settled down to a session of fast gunning. His first toll was composed of thirteen canvasbacks. He bagged twelve of them with his five-shot repeater (though most of his colleagues preferred to have two such guns at hand for fear they would not be able to reload fast enough). In an awesome shooting exhibition, he then proceeded to take one hundred and fifty-five birds between three o'clock and sundown.

The slide-action and semiautomatic guns in use today are little different from those early repeaters, though ammunition has been vastly improved, and the flyway itself is vastly changed.

Punt guns, though they were of less consequence to the Atlantic Flyway than is generally supposed, came to symbolize man's ingenuity and determination in squandering the irreplaceable. In fairness to those who operated the wildfowl cannons it must be repeated that nature's bounty seemed to them a miraculous cornucopia which could never be depleted. What we look back on as obdurate greed they saw as a puny harvest that barely skimmed the superfluity of fowl. Neither the Connecticut mill hand who gunned on Long Island Sound nor the Virginia tobacco farmer who kept a skiff at Craft's Landing was likely to have heard of the Swamp Act or any other land-draining measure that might some day dry up his source of game. The only danger the punt gunner perceived was the very real danger to himself when firing his sometimes unreliable weapon or manipulating an unseaworthy craft that was hardly more than a floating gun carriage.

The punt gun—or simply "big gun," as it was called on the bays where it became most common—can be seen as a logical outgrowth of long-fowler design. There is no evidence that hunters in England or America perceived anything unnatural in using a boat-mounted smoothbore whose great size, weight, and recoil prevented it from being held to the shoulder.

Regarding its original purpose there is some uncertainty. It was probably first used as a harvesting tool by English farmers rather than as a novel sporting arm for the gentry, though it quickly became that, too. By the 1820s, wildfowlers on the English coasts, particularly in Hampshire and Sussex, had developed water-stalking techniques for bagging widgeon, teal, and other water birds with oversized shotguns fired from small boats. The celebrated English shooter Colonel Peter Hawker (and in Ireland Sir Payne Gallawey) awakened interest in the technique as a sport. Hawker contributed improvements to the design of the punt, or sneak skiff, and to the gun as well.

Hawker's own punt gun, which has been preserved by The Wildfowlers' Association of Great Britain and Ireland, is a double-barreled monster weighing one hundred and ninety-three pounds. Its barrels are over eight feet long, have a bore diameter of one and a half inches, and are designed to throw a pound of shot per barrel. The one on the right is a flintlock and that on the left employs the percussion system (a not uncommon combination during the brief period before shooters gained complete confidence in caplock ignition).

Most punt guns were single-barreled, however, and a few of the early ones—weighing a mere thirty to fifty pounds—were occasionally rested on a forked support and fired from the shoulder. As time went on, the trend was toward larger size: a barrel seven to nine feet long, an overall length of ten or twelve feet, a bore diameter of one and a quarter to two and a half inches. Every gun performed best with a particular proportion of powder to shot which had to be discovered by trial and error—regardless of the maker's recommendations. Some were designed for a load of as much as two and a half pounds of shot, propelled by five ounces of black powder. The average shot charge, at least along the Atlantic Flyway, appears to have been two pounds. Pellet size varied greatly, but surprisingly small sizes were popular since the object was to spew forth a dense though wide pattern.

The use of punt guns spread fairly quickly from England to America's East Coast, where it reached its zenith (mostly but not exclusively for market shooting) at the end of the nineteenth century.

The weapons were mounted in small skiffs, usually by means of stanchions and ropes although a few employed swivels resembling giant oarlocks. The stanchions and ropes were arranged to accommodate tremendous recoil, often in such a manner that the gun's butt slammed rearward into a special well where it compressed bags of sawdust, pine needles, sea oats, or other cushioning materials. Springs were also occasionally used, and a kickboard was built into the skiff to prevent recoil from knocking the ribs apart. But regardless of these safety precautions there were authenticated calamities when a gun either blew up upon firing or ripped right through the stern.

Many kinds of boats were used—including canoes and light rowboats in this country—but the most common type on the Atlantic Flyway was a sixteen- to eighteen-foot double-ender skiff with a shallow draft to negotiate the mud flats and a very low silhouette to facilitate sneaking up on rafts of ducks or geese. In addition to the double-ender, which was paddled or poled, there was a square-back type with a hole in the stern through which a sculling oar protruded for use both as a means of propulsion and a rudder. In England there were also larger two-man skiffs. Both partners lay down, and the gunner held his head just high enough so that he could see to give directional signals to the oarsman with a nudge of his foot.

Regardless of the craft's size, the gun was mounted low, with its muzzle just over the bow, and since the weapon was barely maneuverable, the boat itself was aimed. A double-ender was poled or paddled to the general vicinity of the ducks, after which the gunner lay low and maneuvered toward the quarry with small "hand paddles" which resembled slightly elongated table-tennis paddles. When the time came to handle the gun, he could let go of these paddles because they were secured to him with wrist thongs.

The gunner was often chilled to the bone, wet with spray, cramped, stiff, but the sight of a huge "company" of ducks, as the English phrased it, released the glow of adrenalin he needed to paddle forward slowly and silently until the prey was within seventy yards or less. He had been here before, had seen the ducks feeding along the mud flat and planned the stalk. He had waited for several days and nights for the breeze to abate; so much as a slight swell on the water would interfere with aiming. Now, with the boat drifting closer to the ducks, the crucial moment was at hand. He must not make a sound until the instant of the flush. Arms historian Harold L. Peterson succinctly described just such a situation in *Pageant of the Gun:*

*Firelighting for deer as well as waterfowl was common practice.*

Then, with his finger on the trigger, he carefully aimed the gun, elevating its muzzle slightly. He gave a shrill whistle and the startled ducks sprang up in a cloud of whirring wings. Just as they cleared the mud, caught in their most vulnerable position, the big gun boomed. Its shot tore a path through the black mass, and birds dropped like hail.

Assuming that the gunner's aim had been true, he expected to pick up a score of ducks after that pull of the trigger, and perhaps forty or more if he had been lucky. Although the usual rule on the Atlantic Flyway was one man to a boat, sometimes as many as four boats worked together. At the right moment, the leader of the expedition tapped his gun barrel with a coin or other piece of metal to alert his comrades for the flush, and all would fire in unison. A Chesapeake Bayman named Ray Todd told of such a hunt, conducted one evening when there were so many redhead ducks about that "there was a solid roar like thunder every time they flew." It was not quite dark yet, but the ducks were mostly young ones that had not been shot at, and they swam directly at the boats, "so thick you couldn't place a feather between them."

Todd and one partner were each loaded up with two pounds of Number 6 shot, while the other two guns were loaded with a pound and a half. One of the men was deaf, but such problems rarely impeded a waterman—he fired the instant he saw the flash of the other guns. The four cannons boomed almost simultaneously, and then each boat made three trips to shore, loaded to the gunwales with birds—a total of four hundred and nineteen redheads. The performance was repeated three times that night, and by morning the men had killed over a thousand ducks. That was the best night's work Todd could remember. The birds brought $3.50 a pair in Baltimore.

For this kind of punt gunning a low-intensity light was usually employed. The Indians had hunted both waterfowl and big game by firelighting. A small fire of bark was built atop stone slabs on the bow of a canoe, with a large stone or wood slab behind this torch to act as a reflector and hide the hunter's silhouette. The canoe then drifted or was paddled close to the quarry. The light was reflected by the eyes and gleaming feathers of hundreds of immobile ducks or geese, transfixed just as wild creatures today are entranced by the beams of an automobile's headlights. Sometimes, when the birds had just arrived from their isolated breeding haunts, it was as easy as slaughtering penned animals.

The settlers borrowed the idea and improved on it, first using beeswax and tallow candles, then spermaceti candles, then kerosene lamps, and, before the era came to an end, automobile headlamps connected to batteries. The gunners learned to use reflectors that would keep the beam dim and diffuse—the mercury reflectors of tractor lights seemed especially effective for swans and geese as well as ducks. Good results could also be obtained with an ordinary flashlight whose lens was screened with aquatic grass. The light was just strong enough so that the gunner could make out the rafted silhouettes of the birds, never strong enough to frighten the quarry into premature flight.

Yet the birds did learn to avoid areas where lights came on at night and big guns resounded. This being the case, firelighting and punt gunning quickly earned the enmity of hunters who relied on other methods. When Virginia legislated against the lights and the big guns in 1792, Maryland and North Carolina had already banned firelighting, but no one took such laws very seriously. It was not law but the difficulty and hazard of punt gunning that persuaded many watermen to give it up.

A man often had to go far out across ten-foot depths to reach the rafted diving ducks. A sudden wind could swamp his frail craft, or an unseen ice floe could strike. More than one experienced waterman realized too late that he was hopelessly lost at night on a big bay, and his weak gunning light could not begin to reveal the nearest landmark. The possibility that a gun might some day blow up must have been less terrifying than the contemplation of death by exposure or drowning.

The punt gun presented other problems. It could not be left loaded for more than two or three days or the powder would absorb moisture. In England, where the difficulty of the water-stalk and the hazards made punt gunning an appealing sport, costly cartridge-firing breechloaders appeared. A few English punt guns were brought to the Atlantic Flyway, but few if any breechloaders were ever used here. Most of the punt guns in America were less costly models made in New York or Washington, and later in the "outlaw years" there were a few homemade "pipe guns." It was a chore to load a big gun from the muzzle, even on land, and it was said that a punt gunner could be recognized by

*Officer of Fish and Wildlife Service examines gigantic rack of old punt guns and battery guns, confiscated from market shooters.*

the facial burns and scars of past mistakes. Old rope had to be teased apart to make the oakum which was used with cork or paper for the wadding. With the primer tube open so that all the air could escape, two waddings were rammed down over the powder; then the shot was poured in and a final loose wadding added. Packing it too hard would tighten up the pattern and increase recoil. Using powder of too fine a grain could raise chamber pressures dangerously. Too many mistakes could be made too easily. And the waddings were once a common starter of marsh fires.

After making a first shot, reloading on the water was extremely difficult—and extremely dangerous if the gunner failed to swab the last smouldering bit of wadding from the bore. Special tools were invented to ease the task, and there were "night cartridges," made up in advance and encased in fast-burning silk for quicker loading from the muzzle. But the casings interfered with proper ramming, fast ignition, and uniform shot patterns. Moreover, many of the ducks brought down were not killed instantly; they had to be chased and shot again with a shoulder arm.

After years of steady wildfowl decline, federal legislation against punt gunning and other heedless abuses of nature was inaugurated in 1916 and culminated in the Migratory Bird Treaty of 1918, which ended market hunting in America except for sporadic outlawry. It is true that one night during the Great Depression half a dozen firelighters, using three kerosene lamps, killed a flock of forty-seven swans; at $1.50 a bird, they considered the profit excellent, and a customer with a hungry family was unlikely to complain about getting that much meat for a dollar and a half. But such violations of the law were becoming rare before the country was plunged into depression, and not even the drive of hunger could revive the use of punt guns.

In England, the Protection of Wild Birds Act of 1954 limited the size of bores to an inch and three-quarters but permitted the sport of punt gunning to continue. Modern punt guns, made by Thomas Bland of London, are efficient screw-breech artillery pieces and some are double-barreled. Both barrels of the double models fire at once with the pull of a lanyard. Though such an arm may be an intriguing curiosity, the American sportsman does not regret its absence from these shores.

After market shooting was banned, there was a short period when battery guns were used by some of the more recalcitrant purveyors of game along the Atlantic Flyway. The size of punt guns made them difficult to conceal. Many were confiscated, and a man who was caught using one faced the possibility of a heavy fine or (in the case of an incorrigible offender) a prison sentence. The battery gun was a smaller weapon with three, four, or as many as a dozen barrels of moderate dimensions. It achieved pretty much the same effect as a punt gun and was easier to transport or conceal.

Following some early experimentation with individual guns wired together, a fairly typical battery gun emerged. It consisted of the stock-less barrels of old muzzle-loaders mounted side by side in a large, flat block of wood or cement. Two flash holes were drilled at the chamber end of each barrel, the hole on one side serving as a receiving vent for the primer flash and on the other side to send the flash onward to the next barrel. A shallow trench extended across the mounting block from hole to hole, and a few grains of fine powder sprinkled in this trench would carry ignition along to each barrel. A canvas or leather trench cover was intended to protect the shooter from the hot gas and sparks of the explosion. A single hammer and trigger were mounted to the barrel at one side or the other of the assembly, and a single pull fired all the charges—in series, but almost simultaneously.

The barrels were mounted close together to facilitate travel of the flash from one to the next, but were slightly fanned in order to spread the shot over as wide an area as possible. Some employed three 2-gauge barrels, others used four 4-gauge barrels, and there were many that had a larger number of smaller bores. A surviving specimen from the Chesapeake Bay area has a dozen 12-gauge barrels.

The pattern of an average battery gun, with perhaps four barrels, would spread out to cover about ten feet at thirty yards—almost twice as wide a pattern as that of most punt guns. However, ignition was unreliable even on the late-vintage "improved" models which replaced the flash trench with threaded connecting tubes, usually fashioned from an old .22 rifle barrel. Hangfires and misfires were common, as were serious injuries from flashes or explosions caused by faulty parts, loose connections, or overloading.

Battery guns did not appear spontaneously upon the demise of punt guns; multiple-barreled fowling pieces had been tried during the nineteenth century. And later, when federal regulation was answered by a small surge of defiant outlawry, a few double-decked battery guns were built. It is said that in Mexico there were also triple-decked land batteries called "armadas." A few multiple-barreled or oversized guns may still be in use in Mexico, where poaching keeps pace with hunger.

Here, however, such arms have been out of operation since the end of the Depression. Strengthened law enforcement, welcome though it was, probably had a smaller deterrent effect than the danger of firing such weapons. To do so repeatedly, a man had to be very foolhardy or very desperate to make a living.

Antiquarians occasionally have been led astray by the use of the term "battery," which did not always apply to firearms. The sinkbox was most widely known as a "battery" or "battery box" during the early days of its development. This use of the word is of uncertain derivation, yet the term has a felicitous ring. For one thing, the word battery has long been employed to mean a position of readiness for firing: "The gun has been returned to battery after the first shot." For another, the same word has long denoted an artillery emplacement, not necessarily of more than one gun. And finally, the term was and is current among English and Scottish grouse shooters to mean a low turf blind.

But the invention of the battery box—neither a boat nor a floating blind in the usual sense but a flat raft with a central depression that conceals all or most of the shooter's body below water level—appears to have been American. It probably originated on the New York waters where it was first banned in 1839, and where the ban was first openly defied. Today's federal regulations include the decree that "Migratory game birds may not be taken from or by means, aid, or use of a sinkbox," and a sinkbox is defined as "a raft or any type of low floating device having a depression which affords the hunter a means of concealing himself below the water."

There is a fairly common misconception that this prohibition and most of the other bans on techniques and devices deemed abusive (oversized guns, live decoys, baiting, and the like) date from the 1916 British-American proclamation for the protection of migratory birds in the United States and Canada, or from the 1918 Migratory Bird Treaty Act which implemented that proclamation.

The stipulations of the proclamation—or Convention, as it was called —and the treaty established closed seasons in spring and summer, limited the open seasons to periods not exceeding three and a half

*Arthur B. Frost's 1895 portfolio, "Shooting Pictures," portrayed sinkboxes . . .*

months, gave full protection to nongame migratory birds, outlawed the taking of nests or eggs except under special authorization for propagation or scientific purposes, and provided special measures (refuges in some cases, five- or ten-year closed terms in others) for declining species such as wood ducks, eiders, band-tailed pigeons, cranes, swans, and most shorebirds. In addition, such copious restrictions were placed on the traffic in game birds as to effectively end legal market hunting.

Many other federal regulations, which (though they are now all but taken for granted) have been essential in restoring game birds and their habitat, were not written in 1918 but much later. Yet it was the Migratory Bird Treaty Act which made possible these extensions of federal control by empowering the Secretary of Agriculture henceforth to adopt further regulations in accordance with current conditions and studies. (Later, the responsibility was transferred to the Secretary of the Interior, whose department includes the Bureau of Sport Fisheries and Wildlife.) The Treaty Act also authorized appropriations from the Treasury to finance its provisions, and established a penalty system of fines or imprisonment for violations—together with means of enforcement which included federally employed wardens and investigators.

Regulations were added as their need was perceived, and a number were promulgated in the '30s. Some individual states had by then forbidden the use of sinkboxes but they were not universally proscribed until the time of the Great Duck Depression which occurred simultaneously with the nation's economic slough of despondency.

An attempt had been made in 1918, when the Migratory Bird Treaty was signed, to eliminate the sinkbox by prohibiting the shooting of ducks from any boat propelled otherwise than by oars or paddle. It is, of course, permissible to travel to and from a shooting site by motor-driven conveyance, sailboat, or any other craft, and there is no restriction on the type of vessel used to set out and pick up decoys or pick up downed birds. But if the actual shooting is done from a boat, it must be

*. . . decoy rigs, and boat-blinds then in great favor on American waters.*

a manually driven craft. Theoretically, the sinkbox did not qualify because it had to be towed. This, however, proved to be an easily surmountable obstacle for the determined battery shooter. A warden had little basis for arresting a man in a sinkbox fitted out with oarlocks, even if it seemed doubtful that the craft could be rowed far by anyone of less than Herculean strength. In 1927 it became unlawful to use batteries on inland waters and finally, in 1935, the Secretary of the Interior banned all sinkboxes whether oared or oarless.

During the early 1800s these contrivances evidently were used chiefly for waylaying coot and broadbills on the northern bays, and "battery shooting" remained a traditional technique on New York waters throughout the century. In the writing of a splendid 1937 book entitled *Gunner's Dawn,* that brilliant painter, writer, and outdoorsman Roland Clark consulted the entries for December 18–21, 1889, in his battered old gunning diary, recording a Long Island hunt with his two brothers and a guide named Captain Robins. Clark and his brothers drew straws for the use of a "double battery" (two-man sinkbox) on Great South Bay near the Fire Island lighthouse. He felt a pang of disappointment upon drawing the short straw, which relegated him to a pond blind on an icy Bay Shore marsh. But in spite of snow flurries and a shifting wind, he managed to kill seventeen black ducks and a widgeon drake by noon. "I never did confess how many shells I used on those eighteen ducks," he said. "It's something of a secret I have carried all these years." His brothers meanwhile killed thirty-three ducks, mostly redheads, from their offshore battery and they were not especially proud of the score, which would have been higher if the wind had been less changeable. Remembering how he had crouched shivering in his blind and how his brothers had huddled in a cramped battery (likened by some sinkbox shooters to a medieval instrument of torture), Clark observed that "surely the congenital duck-shooter is imbued with undying patience." Such behavior might be mystifying to the uninitiated but not to an addict like Clark, who described the rapture of that winter morning, "with a rose light breaking in the east and dark lines of ducks in silhouette."

From such reminiscences, the sinkbox emerges not just as a market gunner's tool but as a sportsman's delight and affliction. If credit for its origin must go to New York, it achieved its greatest popularity and refinement of design on the Susquehanna Flats, near Havre de Grace, Maryland, where armies of ducks and geese spew from the flightlanes over the Delaware and Susquehanna rivers onto the broad Chesapeake waters. Countless thousands of ducks once wintered on this fifteen-square-mile delta, enriched by these major rivers and many smaller ones—the Elk, the Bohemia, the North East, and more. Some authorities say that greater numbers of canvasbacks gathered here than anywhere else on earth. And they also say that more cans, redheads, and scaup were bagged from Atlantic Flyway sinkboxes than by any other means anywhere.

There were two basic types of sinkbox, the "sit-down" and the "lay-down" design. The sit-down battery had a sufficiently deep compartment so that a seated shooter could hunch down until only his head, or head and shoulders, might be seen above water. Batteries were invariably surrounded by tremendous rafts of decoys—from a hundred to the "customary" three hundred, or sometimes more than six hundred. Among all those blocks, a man's head was not likely to be noticed by incoming flights of birds.

The lay-down battery had a shallower draft, permitting the hunter to remain concealed below the water level only by lying supine. This design required less ballast than a sit-down box to make it ride low enough in the water, but its exquisite discomfort rendered it less popular among wealthy urban sportsmen who expected to be pampered. The difficulty entailed more than being cramped or enduring a stiff back. As wryly noted in Raymond Camp's *The Hunter's Encyclopedia,* "the shooter employed his back and stomach muscles to bring him erect for the shots, and after about twenty such movements the novice shooter became convinced that he had developed a severe hernia."

When battery shooting became a fashionable sport late in the nineteenth century, it was customary for a guide to shelter visiting hunters on a houseboat which also served to hold great heaps and barrels of ducks until they could be taken ashore. The men ate and slept aboard, and they were ferried out to the anchored sinkboxes in a "pick-up" boat that was also used to retrieve ducks. A good season attracted more hunters than there were sinkboxes, but the sportsmen did not mind waiting their turn in the convivial warmth of a houseboat, playing cards, eating, drinking, napping, trading cigars and tall tales of tall shots. On a cold day along the Susquehanna Flats, where lay-down boxes predominated for a long period, an average sportsman could endure no more than about an hour in a battery. Then he would hold his gun up as a signal—or supplication—for the pick-up man to retrieve him and let some other madman have his chance.

There were double as well as single batteries, and even some four-man sinkboxes. Such arrangements must have afforded substantial savings in time and money for the operators while providing the gunners with the company which misery is said to love. Since the total cost of a houseboat, pick-up boat, battery, decoys, and gear might be eight or ten thousand dollars, such operations had to bring in good profits. And no doubt they did. In 1952 the Baltimore *Sunday Sun* printed a brief memoir by Bennett Keen, a former market gunner and guide who recalled that his brother John, who had also guided sinkbox sportsmen, had counted Grover Cleveland and J. Pierpont Morgan among his clients.

So lethal was sinkbox shooting, he declared, that in the days before the limits, a market shooter could often kill over a hundred ducks in a morning, and his brother had once taken more than three hundred and fifty.

In 1920, according to Keen, "at the height of the sportsman's use of the Flats," there were some fifty sinkbox rigs at Havre de Grace, and most of these consisted of one large boat, two pick-up boats, and a battery. "The sinkbox itself was shaped mighty like a coffin, and was just about as deep."

During World War II, the bodies of two English seamen were washed up on Ocracoke Island in North Carolina. That state had passed its own law against sinkboxes in 1935, the same year they were banned by federal regulation. The island's last two batteries were used as coffins for the drowned sailors. The contours of the boxes required no alteration.

Hinged to the top of the compartment of a typical sinkbox were wooden and canvas or burlap wings, riding flat on all sides to absorb any chop or swells. Large weights, ranging up to forty pounds each, took the craft down almost flush with the water. Even this ballast often would have been insufficient without cast-iron or lead decoys, some weighing twenty-five pounds apiece, mounted on the platforms near the compartment. Farther out on the canvas portion of the wings rode flat-bottomed wooden decoys, weighted at the base and anchored by lines

125

that passed through openings in the canvas. Battery decoys have become rare collector's items and, although the wooden ones were frequently old standard blocks with the bottoms sawed off, some were minor works of art.

Theoretically, a sinkbox was reasonably safe in breezes of up to fifteen knots, but if the water became any rougher there was danger of swamping. So many accidents occurred that one is led to question the safety of these floating coffins even in a calm. Most of the drownings occurred when a battery simply swamped in the eight- or ten-foot water where the hunting for diving ducks was best, but there was at least one recorded tragedy when a gunner inadvertently fired his gun, blew a hole in the floorboards, and quickly submerged.

The sit-down battery appears to have been a much later development than the lay-down type, and some authorities believe it originated in the Carolinas. Harry Walsh sought firsthand information about its origin from an old market gunner named Luther Parker, of Knotts Island, North Carolina. Parker said that the first sit-down box he had ever seen—and he believed it was the first one made, or one of the first—was designed by Charlie Balance of Bell's Island, North Carolina. The slight extra comfort of such boxes was offset by danger, for as Parker added, they were even heavier and harder to handle than lay-down batteries. He also noted that batteries of both kinds probably added to the popularity of oversized decoys. One winter, the "Ocracoke boys" came up to Currituck Sound, where most of the fowl were gathered, and they brought along their large brant decoys, which they painted like redheads. After that, Parker and his colleagues switched to oversized decoys because they were visible to birds at great distances and also helped to hide the batteries.

The combination of batteries with big spreads of big decoys was so lethal that many men were willing to face the hazards for a number of years. Observers generally agree, however, that the danger of sinkbox shooting caused batteries to fall into disrepute before the law consigned them to permanent oblivion.

Several other wildfowling vessels which were developed in the nineteenth century were boats in the true sense, and if they were—and are—less dangerous and less colorful than the sinkbox, they are hardly less ingenious in their special-purpose designs.

The sneakbox (more properly called the Barnegat Bay sneakbox in tribute to its ancestral home) is so much nobler a craft than the sinkbox that writers who confuse the terms elicit gasps from fastidious sportsmen. That portly but tireless wildfowler Stephen Grover Cleveland

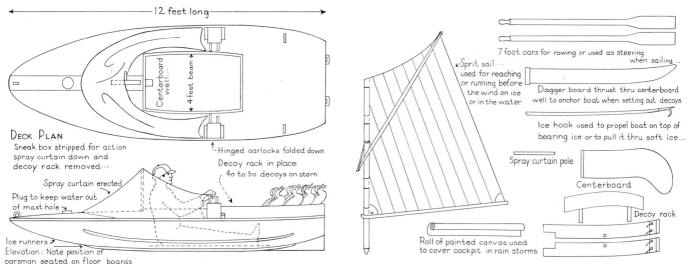

**12 feet long**

DECK PLAN
Sneak box stripped for action
spray curtain down and
decoy rack removed···

Centerboard well···

4 feet beam

Spray curtain erected

Plug to keep water out
of mast hole

Ice runners
Elevation: Note position of
oarsman seated on floor boards
and fully protected from wind
and spray···

Hinged oarlocks folded down
Decoy rack in place
4o to 5o decoys on stern

Sprit sail···
used for reaching
or running before
the wind on ice
or in the water

7 foot oars for rowing or used as steering
when sailing···

Dagger board thrust thru centerboard
well to anchor boat when setting out decoys

Ice hook used to propel boat on top of
bearing ice or to pull it thru soft ice···

Spray curtain pole

Centerboard

Decoy rack

Roll of painted canvas used
to cover cockpit in rain storms

*Classic duck boats are still built from plan of Barnegat sneakbox which appeared in* Field & Stream *in 1942 and again in 1971.*

(who also achieved some prestige as a politician by ascending twice to the presidency of the United States) shot birds from both the sinkbox and the sneakbox. He undoubtedly settled his bulk a good deal more comfortably in the latter than in the former, and at about the turn of the century he enjoyed sallying from his Princeton home to Barnegat, the fine coastal area below Bay Head and above Cape May, New Jersey.

At about the same period or a little later Van Campen Heilner, who was then associate editor of *Field & Stream,* wrote about his good takes of cans, broadbills, and redheads, shot from a camouflaged sneakbox on Barnegat Bay.

So trim and seaworthy was the standard twelve-foot sneakbox that it was used in club races, and Nathaniel Holmes Bishop boarded his custom-built $75 craft to drift, sail, and row twenty-six hundred miles down the Ohio and Mississippi, from Pittsburgh to New Orleans, afterward describing his adventure in a book entitled *Four Months in a Sneak Box.* Although such a craft was only about four feet wide and a foot deep amidships, and was decked over except for a center cockpit, there was room enough below decks so that the market gunners who perfected the design could stow oars, pole, paddle, ice hook, shells, two guns, foul-weather gear, extra clothing, lunch and a bottle, and sometimes a sprit sail, detachable rudder, tiller, centerboard, and "dagger board." The last-named piece of equipment is a blade-shaped board about four feet long which is lowered through the centerboard well and down into the mud to hold the boat stationary while the shooter is setting out decoys.

More or less standard gear also includes a detachable wooden rack to keep stacked decoys from spilling off the aft deck, a crown about four inches high around the cockpit to shed water, folding oarlocks (raised above the cockpit crown on wooden blocks), and a canvas apron that can be erected as a spray curtain forward of the cockpit and sloping down along the sides to the oarlocks. Some models even boast hinged hatches which can be locked down over the cockpit for off-season storage of gear.

The shape of the square-sterned, round-nosed, spoon-bottomed Barnegat sneakbox has been likened to a pumpkin seed, but there have been many minor variations. On some the bow was sharpened where it cleaved the water to reduce noise; on others, sheer sides stiffened the structure to provide better shooting stability. However, the basic design was so reliable, maneuverable, and versatile that severe modifications were uncommon except for the sail-racing sneakbox which evolved from the hunting boat. William N. Wallace considered the racing version (which is still favored by Barnegat boating enthusiasts) sleek enough to warrant inclusion in his *Macmillan Book of Boating.*

The Barnegat Bay Sneakbox was an early one-design class. It evolved from the early nineteenth-century sneakboxes used for shallow-water gunning. . . . J. Howard Perrine produced a formal design in the early 1900's, a 15-foot, gaff-rigged centerboard sloop that weighed only 350 pounds and could sail in as little as six inches of water. . . . Over 3,000 boats of this type were sold. . . . The successor boat, the 14-foot 8-Ball Sneakbox, is a modern one-design class on New Jersey's Barnegat Bay.

The original shoal-draft sneakbox can be anchored as a floating blind, camouflaged with sedge to look like a drifting hummock and surrounded with decoys. Or, since its freeboard rises only a few inches above the surface, it can be used to jump-shoot ducks in a marsh or even sneak up on big rafts of them out on open water. Its stability renders it safe under a bay wind, and oak or iron runners can be mounted under the hull if a gunner wishes to sled across an icy marsh and into a stand of reeds where the boat becomes a blind.

Harry Megargee contributed construction drawings of the classic sneakboat to a 1942 issue of *Field & Stream* in which he wrote:

*Typical modern duck boat rides low and provides ample deck space for gear.*

*Before streamlining evolved, square-ended hunting skiffs were common.*

I have seen the air alive with ducks on Chesapeake Bay, when golden opportunity went begging because of inability to get off to the "booby blind" against the gale. . . . On such a day, a Barnegat bayman would have taken his sneak box offshore without turning a hair or shipping a cupful of water, set out his decoys, and killed his limit of ducks.

The magazine still receives occasional requests for Megargee's construction plans, and Barnegat Baymen continue to shoot ducks and brant from classic sneakboxes.

Another type of spoon-bottomed boat developed during the nineteenth century is the famous Merrymeeting Bay scull, probably originated by market gunners and later a great favorite of Maine sportsmen. It is a sharp-bowed, square-sterned craft with a long cockpit that can accommodate two men and their gear quite comfortably. The cockpit extends all the way back to a stern seat, with only narrow decking at the sides and a little forward deck. The boat can be poled or propelled by a curved sculling oar which passes through a leather or rubber sleeve lining a hole in the transom. A man can sit or kneel in the stern while sculling, and with a little practice can lie supine as he sculls silently toward the quarry. Like the sneakbox, the Merrymeeting scull rides low in the water, is easily grassed over for camouflage, and remains in use today.

So, to a lesser degree, does the Connecticut scull, a V-bottomed variation with a one-man cockpit. Since the Connecticut boat is much the more seaworthy of the two, it is sometimes painted dark brown, draped with rockweed, and taken offshore to outlying reefs. On the other hand, its V-shaped hull prevents it from being used in winter for "ice sculling," a technique for which the oval-bottomed Maine boat is sometimes used. The advantages and disadvantages perhaps cancel one another out. Sculling boats of these and other, slightly modified, persuasions accounted for tremendous kills by the northern market gunners.

Extant records include some astounding one-shot kills with large-bore double shotguns fired from sculls: nine geese, for example, dropped by both barrels at close to seventy-five yards.

For marsh shooting, however, flat-bottomed skiffs have probably always been most popular. If a well-made sneakbox can skim along in four inches of water, a light skiff can often work through areas of less depth, more grass and mud. Until the mid-nineteenth century, the Atlantic Flyway skiffs were as square-ended as scows, front and back. It may be that design was ultimately influenced by the maneuverability of canoes, whaling boats, New England dories, or certain of the Maine guide boats used chiefly for fishing and deer hunting. Whatever served as inspiration, the skiffs became trim double-enders until the advent of the outboard motor spawned the square-back design. Such boats have remained in favor since they provide motorized transportation over deep waters and can be rowed, poled, or paddled over the shallows. And although the skiff was originally intended for marsh shooting, it is sufficiently seaworthy for reef hunting.

It was from the simple flat-bottomed skiff that the sleek Long Island scooter evolved. A scooter remains fundamentally a skiff but with special equipment: a sprit sail, small grapnel, scooter hook, and bottom runners. Rigged to a long line, the grapnel can be thrown out on thin ice. The scooter hook is still handier. It is like an ordinary boat hook, except that the hook combines a pike with a palm shaped like a pointed hoe. The pike can be used for poling across ice, the palm for digging into the ice and pulling the boat up on it. As for the sail and runners, with a stiff breeze they can take the boat over the ice at sixty miles an hour.

Various other "grass boats," whose use has spread westward from the Atlantic Flyway, are made in both double-ender and square-sterned models which look like hybrids of the skiff and the scull.

The rail boat, too, is a skiff, usually with a flat, smooth bottom and extremely shallow draft for poling over the wild-rice flats frequented by marsh hens. As a rule, the pusher stands on a platform resembling an oversized stern seat, and his pole may have short, spread legs or a blade since an ordinary dowel-shaped pole can penetrate deep into the mud and stick there instead of pushing the boat along smoothly. Some rail boats, especially in the South, are equipped with a high, round-seated kitchen stool secured approximately amidships so that the gunner can perch comfortably. Most of the northern boats have conventional seats, but the Connecticut gunner traditionally stands as he awaits the flush of birds.

RAIL SHOOTING.

Evolution of classic
shallow-draft skiffs for
rail shooting or
poling over duck shoals
can be seen in two
19th-century prints at
left and in 1931
painting, "At Sundown,"
by Frank W. Benson.

Most Connecticut rail boats also have a bit more freeboard than their southern counterparts, and some of the older ones are nearly double-enders, with a very narrow, notched transom to accommodate a sculling oar. Most of the modern ones, both in the North and South, share the skiff's square stern so that the party can be motored to the flats and only then does the pusher begin his back-straining labors with the pole.

Shooting rail (or, for that matter, snipe and various shorebirds) has a certain upland-gunning quality about it. Since rail boats are used to push through tall growth and flush birds that are reluctant to rise, concealment or disguise is not an important consideration. But it has always been a prime consideration with duck boats.

Famous Thomas Eakins watercolor "Whistling for Plover" (above) and 1869 Currier & Ives print "Beach Snipe Shooting" (right) prove how easily some migratory birds were lured into range, but another Currier & Ives, lithographed from 1862 oil by Arthur F. Tait, shows pigeon shooters in heavily brushed blind, working movable live-decoy rig.

132

A good duck boat—that is, a *hunting* boat, as opposed to a vessel intended only to carry the gunner and his gear to a shooting locale—must serve both as a water vehicle and as a blind or foundation for a blind. The sneakbox and the scull rise only a little higher from the surface than the platform wings of an old-time sinkbox, and the lay-out boat, that ultramodern fiberglass-decked counterpart of the traditional Maine scull, usually has an even lower silhouette than that of its classic forebear.

The blind has always been of great importance in migratory-bird hunting of most kinds along the Atlantic Flyway. Sometimes the blind may be rudimentary or it may be no more than a natural hiding place. A dove hunter can find opportunity to miss a plethora of targets while crouching under a hedgerow at the edge of a feeding field. And it is true that a connoisseur of hunting art may come across an 1869 Currier & Ives lithograph of a snipe shooter, sitting completely exposed on an open beach as birds fly down to inspect his stick-up decoys, or an 1874 Eakins painting of men squatting unconcealed on an open flat and merely whistling to entice yellow-legged plover into range. But it is equally true that the most unsophisticated birds—species which have seldom been sought by man in a particular area or, conversely, species almost contemptuously accustomed to man's proximity—quickly learn to avoid the gun. And the connoisseur may also come across an 1862 Currier & Ives lithograph of Arthur F. Tait's painting, *Pigeon Shooting: Playing the Decoy,* in which two gunners crouch in a blind of thickly interwoven branches and foliage while using cords to manipulate tethered and mounted birds on a manmade roosting tree.

Though hunting did not become a popular theme of American artists until the mid-nineteenth century, there are earlier depictions, both European and American, of men using brush, reeds, and other materials as blinds. And written references to ''stands'' and ''huts'' and ''places of concealment'' date back to remote antiquity, so it can be assumed that the Baron Lahontan was not unduly surprised to find the Indians shooting from ''huts upon water'' at Lake Champlain in 1687.

Two and a half centuries later, when sinkboxes were prohibited, some of the gunners on the Susquehanna Flats devised a new sport called ''body-booting.'' Body-boots, much like today's chest-high waders, were worn by fishermen hauling seine. Duck hunters who wore them could remain reasonably dry in at least four feet of water. A pick-up boat brought the gunners out to a bar, put them off, surrounded them with decoys, and then withdrew so as not to alarm the ducks. Of course, this

substitution for sinkbox shooting would have been cold comfort even without the difficulty of keeping one's shotgun and ammunition dry, so the gunners soon returned to traditional boats and blinds.

Wildfowlers have long known that the best blind is not a blind in the usual sense but a natural hiding place—tall reeds, bushes, driftwood, a clump of boulders—where birds frequently fly low or alight. The less artificial concealment or disguise is needed, the more likely ducks are to make an unsuspecting approach. In some of the finest shooting areas, nature has not provided sufficient concealment, and artifice is a necessity; in others, concealment abounds, and sometimes man has been content to use what was offered, but more often he has succumbed to the temptation of "improving" his disguise, enhancing his comfort, or merely meddling with nature.

Where Lahontan and his Indians killed their "wonderfull numbers of geese, ducks . . . and an infinity of other waterfowls," men have long shot ducks from rock blinds, cedar blinds, driftwood blinds, houseboat blinds, and other hideaways both primitive and luxurious. About eighty percent of the ducks shot on the Atlantic Flyway are puddle ducks, mostly blacks, mallards, teal, widgeon, and woodies, though not always in that order. But Lake Champlain is a perfect point of concentration for two big-water species, American golden-eyes and scaup (or whistlers and bluebills, as most Vermonters call them).

On November mornings cold enough to make the surface steam and mist, with a southerly wind bringing snow squalls, two gunners make a practice of putting their twenty-foot canoe into the water west of Middlebury and paddling to a slate-covered island a hundred yards offshore. There they set out a couple of dozen whistler decoys, most of the stool directly in front of a big rock pile at the island's edge but with half a dozen blocks angling out on each side to form "hooker lines" that may attract wide-swinging ducks. Thirty yards from the blind the water is at least ten feet deep, and it is therefore a perfect spot to toll in whistlers and bluebills.

Some eighty thousand golden-eyes winter on the Atlantic Flyway, and they have been building up around Lake Champlain since the beginning of November, congregating in large numbers near the mouths of tidal rivers. Curiously segregating themselves by sex, they fly early each morning, following the shorelines but seldom passing over any land. The little island rock pile is therefore an ideal blind when fog or snow nudges the flights down to no more than a dozen yards above the water. The partners usually take their limit.

One of the gunners often occupying this blind is the writer Jerome B. Robinson, who described some of the more celebrated Champlain blinds in the Autumn, 1970, issue of *Vermont Life*. Some are situated on Malletts Bay near the Sandbar Game Management Area (where there are probably more duck blinds than anywhere else in the state). Others are located to the north, on Grand Isle above South Hero. One of the Grand Isle rock blinds, owned in recent years by Dr. John Abajian, accounts for more ducks than any other blind on the vast lake. It has been in use for more than a century and is a veritable fortress of granite weighing several hundred tons, yet it is annually demolished by late-winter ice floes and heavy waves, and painstakingly rebuilt by Dr. Abajian and his friends. Its nooks and boulder turrets can accommodate up to ten guns, and among the lower rocks is an inconspicuous hole from which the dogs can mark fallen ducks.

When a southwest wind has the ducks flying well at this point, waves crash into the boulders, drenching the men's parkas and icing over the rocks. "It is hazardous to stand up," Robinson wrote. And "there is always the chance on a stormy day that a strong wave will knock the heavy granite shooting stations in upon the shooters."

There are smaller, less elaborate rock blinds, and there are "cedar islands" which are actually camouflaged houseboats on pontoons, "warm and comfortable as any deer hunter's mountain camp. A house much larger than any ice-fisherman's shanty squats under a bushy exterior in the center of the craft. Inside is a card table, two or three bunks, a cooking stove, comfortable chairs, gun rack, sink and gas light. Some even have a separate kennel area for the retrievers."

Perhaps the most stylish of all such ducking domiciles was an eight-by-ten-foot shooting cottage, a one-man affair built by another physician, the late Dr. Bradley Jackson, on a shoreline granite ledge at Jackson Point near South Hero. Inside were a stove, bookshelf, gun rack, table, big brass bed, and framed mirror. There was one very large window, hinged at the top and kept open on a shooting day. After lighting the stove, it was the doctor's practice to lie in bed reading and occasionally glancing at the mirror on the wall, which was so hung as to reflect the window view of his rig of decoys.

When ducks dropped in among the decoys, Dr. Jackson put his book aside, took up his shotgun, ambled to the window, and flushed the birds from the water—presumably by shouting, waving, or firing into the air After the dog retrieved his birds, he returned to bed and book.

Houseboat blinds of the barge type were common on this big northern lake during the halcyon days of ducking in the late nineteenth and early twentieth centuries. They have been replaced by a few wide-decked pontoon craft with shooting huts erected upon them. The gunners commute to them by outboard, and most of the shooting is for bluebills—greater scaup.

Both bluebills and whistlers toll to Champlain's rock blinds and tower blinds. The towers, though not so luxurious as houseboats, are comfortable indeed. They are wooden structures built on pilings driven into the lake bottom. The deck is three feet or so above water so that a motorboat may be berthed underneath it. The shooting compartment is generally a roofed plywood structure with large shooting ports but a fair degree of protection from wind, snow, or rain. Inside one often finds benches and a small stove. There may be room for two, three, or six gunners.

Another approach to the attainment of height is Champlain's most ingenious blind, an elevator box that slides up or down a twenty-foot granite bluff at South Hero. Mechanically ornate though it is, cedars easily camouflage it. Lifted by a cable and electric winch, the box slides on rails which also hold a stairway, enabling a retriever to clamber down after ducks and climb up again with his burden. It was still another physician, Dr. Charles Rust, who built this wondrous contraption.

Even more sybaritic compartment blinds have been built by clubs of wealthy sportsmen on the tidewaters of New Jersey, Maryland, Delaware, Virginia, and the Carolinas. They are permanent or semipermanent cottages with wide shooting ports, camouflaged with a thatch of reeds and brush, and they are situated on points overlooking good feeding waters. Some have space for half a dozen men, warmed by electric heaters and entertained by card games or portable television sets while one or two lookouts may watch for flights.

Even when such a blind is partially sunk below ground level, somewhat like a plainsman's sod shanty but with the granite-founded solidity of a baseball dugout, it is a bit conspicuous for duck shooting no matter how ideal for poker. Utility is sacrificed to luxury. A more practical though still more costly variation on the theme is the comfortable marsh-side lodge, with separate shooting annexes: small blinds accommodating two or three men each and situated at a reasonable distance.

Two types of sunken blinds have for generations proved their value. One is the sunken-box blind, most often found on a point, island, or fringe of a cove. It is a one- or two-man wooden box, roofless, with a bench for the gunners and a narrow shelf for ammunition and small accessories. Frequently, thatched screening or some other device to provide camouflage without hindering visibility slides over the top. The other variety is the pit blind, most often used for goose shooting at or near feeding fields. This is a one-man affair which seems to have been in existence at least throughout this century. The pit must be lined if the shooter is not to find himself sitting in water or mud, and for many years the favored lining was a big empty olive cask or pork barrel, satisfactorily watertight and sometimes with a built-in seat and shelf. Water seepage was minimized by dropping in a drainage-layer of rocks before lowering the barrel.

In recent years, the barrel has largely been replaced by a roomy, commercially produced metal drum equipped with an ammunition shelf and a swing-around seat. Between the pit and a circular roof is a fairly wide slit providing a three-hundred-and-sixty-degree view. The roof, camouflaged by reeds, cornstalks, or almost any vegetation—or just mud—pivots horizontally out of the way for a shot. It can be swung off instantly when birds come within range.

Ever since the arrival of the sneakboat, a favorite means of concealment on the shallows of Barnegat Bay has been the cove blind. This is a mud enclosure built in the water near the shore of a cove and large enough to hide one, two, or three sneakboxes. It is fashioned by driving hundreds of fresh green stakes into the mud so that they jut only eight

*Pit blinds of various types have demonstrated their value for many decades.*

137

or ten inches above the water. Mud and grass must be tamped around and over these stakes, which are there only to prevent the earthworks from crumbling away. Camouflage-stained canvas is stretched over the front and rear of each boat, and when grasses and weeds are scattered on this, the appearance from the air is that of a little peninsula in the cove.

A simpler blind, common not only at Barnegat but along the entire coast, is the tule or brush affair which can be erected on any shore by building a frame of lathing, or stakes and chicken wire, thatched and woven with vegetation to look like a clump of natural growth. In one form or another, this type is probably more ancient than any other, with the possible exception of driftwood on beaches or piled rocks on boulder-strewn shores.

Offshore, however, the oldest may well be the stilt, or reef, blind, which is one version or another of the Lake Champlain tower, erected on poles or pilings. A boat often can be concealed beneath the platform, which is reached by ladder and camouflaged with rush, reeds, and so on, woven into any convenient type of frame.

Lower and probably more effective, but useful only in rather shallow water, is the stake blind which was probably used as long ago as the first settlement of the Atlantic Flyway, since structures of this type were then known—as were tule and grass blinds—in Europe. Employing far fewer stakes than are necessary for a cove blind, and permitting them to protrude several feet from the surface, a small boat enclosure is built and camouflaged with woven branches and foliage. To the anserine eye, it appears as a tiny island; to the human eye, it resembles the long, narrow hedgerow blind, either natural or constructed, which is often found in the company of pit blinds along the edges of feeding fields.

Somewhat more ingenious is the float. Since it protrudes substantially above the waterline, it is legal even when it is built around a boat. The camouflage can consist of almost any local vegetation from reeds and grass to wild rice or cattails. More elaborate floats are built upon big rafts secured atop five-gallon metal drums. Sometimes a hut is erected on this platform, and there is room enough to prop up whole evergreen trees as camouflage. Such a manmade island is perhaps not as easily movable as the old sinkbox but is safe, legal, and probably just as effective as the Carolina curtain blinds, those sunken concrete and wood affairs that replaced sinkboxes in the surf of the Outer Banks.

Like the guns and the boats of the Atlantic Flyway, and like the watermen themselves, blinds have slowly acquired a more deadly sophistication in the course of the last two and a half centuries. It is probably safe to claim the same growth of crafty sophistication for most species of migratory birds. As in the case of the Atlantic seaboard's ruffed grouse, which the Puritans were able to club with broomsticks and which have evolved into the wariest of eastern uplanders, survival of the fittest has meant survival of the most distrustful ducks and geese. There has been no way for the birds to defend themselves against loss of habitat or the ravages of pollution or the willful ignorance of game management that long prevailed, but in terms of the hunted and the hunter a remarkable balance has continued to bless the flyway.

*As birds swing over, hunters rise out of haystack blind.*

# Chapter Ten

# A CONSTELLATION OF SWANS

North of the Zodiac, between Pegasus and Lyra, there is a constellation on the Milky Way known as Cygnus, the Swan. To some it resembles nothing more haunting than an open parasol. To navigators concerned only with its brightest stars, it is the Northern Cross. But those whose vision is imaginative can see the parasol top as outstretched wings, the handle as the slender neck of a swan. So it was seen by the myth makers of Greece, who could perceive in the stars the drama of Zeus, god of the elements, father of gods and men, assuming the form of a great swan when he coupled with the wife of a Spartan king and thus begat Pollux and Helen of Troy.

*Cygnus,* the Latin word for swan, derives from the Greek *kyknos,* which is akin to the Sanskrit terms *suci* (white, shining) and *sakuna* (a large bird). The Greeks and Aryans were not unique in their vision of a bird among the stars. The brightest light in the constellation Cygnus is Deneb, a star of the first magnitude centered at the rear of the sparkling wings. Its name is an abbreviated corruption of the Arabic *dhanab aldajajah,* tail of the hen. Hen or swan, it was an enormous bird to the ancients, fixed against the black heavens in eternal, motionless flight. The image remains felicitous; a swan, or brood or flock of swans, is as white and shining in the distance as a constellation of stars.

The body of a swan cloaking the spirit of man has been ubiquitous in legend. A swan, drawing the boat of the medieval German knight Lohengrin to the rescue of Princess Elsa, was afterward revealed in its true form, as Elsa's brother. Teutonic mythology told of swan maidens who transformed themselves into swans by donning a magical swan shift—a garment of white feathers. Similar beings appeared in Japanese myths and in those of some American Indian tribes.

Yet the sight of a constellation of swans is a gift which man has not invariably treasured. Many a market gunner on the Atlantic Flyway considered the swan a plume or meat bird at best, a thief of duck habitat at worst. By 1916, swans were protected on their primary wintering grounds in Maryland, Virginia, and North Carolina, and two years later the Migratory Bird Treaty brought them under complete

# THE BOLD AND LUSTY SWANS

*Gliding regally past shoreline, whistlers seem to want
no company but that of fellow swans, yet when foraging
they mingle with other aristocrats.*

Upper photographs are of mute swans, once "park birds" but now wild and free. Swans in lower photographs are whistlers at Blackwater Refuge.

*Flaming orange bills and strange black forehead knobs distinguish mute swans from native whistlers, but both strains are equally majestic symbols of truly wild wetlands.*

THE HAUNTS OF THE WILD SWAN.
CARROLL ISLAND - CHESAPEAKE BAY

*In 1872 print depicting Carroll Island, on Chesapeake Bay,*
*birds probably include both whistlers and mute swans.*

federal protection, but violations of these laws were common for several years. The reason is evident from Arthur Cleveland Bent's 1925 Smithsonian bulletin, in which he quoted observations of the whistling swan (*Cygnus columbianus*) reported nine years earlier:

The food of the Whistling Swan is largely vegetable, which it obtains by reaching down with its long neck in shallow water, occasionally tipping up with its tail in the air when making an extra long reach. While a flock of swans is feeding in this manner, one or more birds are always on guard watching for approaching dangers, as the feeding birds often keep their heads and necks submerged for long periods. . . . In Back Bay, Virginia, and in Currituck Sound, North Carolina, the Swans feast on the roots of the wild celery and fox-tail grass; they are now so numerous that they do considerable damage by treading great holes in the mud and by rooting and pulling up the celery and grass; they thus waste large quantities of these valuable duck foods, much more than they consume, and consequently spoil some of the best feeding grounds for ducks, much to the disgust of sportsmen in the various clubs, who are not allowed to shoot the swans and have to submit to this interference with their duck shooting. The swans are really such a nuisance in this particular locality that a reasonable amount of shooting might well be allowed; these birds are so wary that there is little danger of any great number being killed.

Today, having been protected for so many years, the swans on Currituck, Back Bay, and the Chesapeake are no longer wary; on the contrary, they are quite bold. They may be seen cruising about, dipping for food, while a little convoy of ducks trails them to pick up leavings plucked from a bottom nearly unreachable for the small dipper species but conveniently afloat in the wake of swans. A controversy over their protected status has continued intermittently ever since the publication of the 1916 studies.

*On secluded creek,
mute swan rises from
her commodious nest.*

There are three representatives of the subfamily *Cygninae* in North America. Two of them, the whistling swan and the trumpeter (*Cygnus. buccinator*), are native to this continent. All three, like the majority of swan species, are white and shining when they reach maturity. The young are tinged with gray, brownish-gray, or yellowish-gray, and yearlings are called "gray birds" in some areas.

The immigrant among the three varieties is the mute swan (*Cygnus olor*), long domesticated in Europe despite a sometimes irascible temperament. Slightly larger than the whistler and even more graceful on the water, it measures nearly five feet long from tail to bill tip when the neck is held straight forward. It rests higher upon the water than the whistler and nearly always swims with its neck curved in a sweeping S and its bill inclined downward, as if it were consciously posing for an eighteenth-century English painter of manicured landscapes.

All swans occasionally swim with their wings raised in an arch over their backs, and it is probable that the females of most species sometimes permit the young to ride upon their backs. However, these habits are most pronounced among mute swans. The mother will lower herself in the water to facilitate boarding, but the old legend that she helps the infants up with one foot is questionable. This belief probably arose from a peculiar habit—shared by whistlers and trumpeters—of swimming with only one foot while the other rests, sole up, along the lower part of the back.

In *The Birds of Norfolk,* a three-volume work of the late nineteenth century by Henry Stevenson and Thomas Southwell, the description of mute swans is summarized in a rhetorical question:

What prettier sight presents itself upon our inland waters with such a group disporting themselves in the bright sunshine of a summer's day, when the pure whiteness of the old birds' feathers contrasts with the green background of reeds and rushes, and the little grey cygnets on their mother's back are peeping with bright bead-like eyes from the shelter of her spotless plumes?

Swans of all kinds have been shot for sport, for feathers, and for food, but in the days of market gunning swan meat brought small profit because only the flesh of the young was considered a delicacy. Mature birds were invariably tough. The reason for domesticating mute swans was almost purely esthetic, so decorative were they on the mirrorlike ponds and streams of great estates. They were brought to America as "park birds," but on both continents they have frequently reverted to the wild state, preferring the vicissitudes of the unkempt marsh to the placidity of cultured waters. Indeed, mute swans are still commonly called wild swans in England, *cygnes sauvages* in France.

Having encountered congenial habitat and immunity from the gun in America, mute swans have taken to breeding as well as wintering on Long Island, along the Jersey coast, and up into the Hudson Valley, where the whistling swan is rare, the trumpeter swan only an echoing memory. And slowly but unquestionably, mute swans are extending their wild range along the Atlantic Flyway. They have spread northward into Massachusetts and southward over Chesapeake Bay.

Like other cygnine birds, they pair for life, and a mature male, or cob, is fiercely, haughtily beautiful when he advances to shield his family from intruders. There are mighty springtime battles among young cobs contesting for females, called pens. Then there is a serene period of nest-building, setting, guarding. Though the female normally attends to incubation, it is said that if she dies her mate will hatch and rear the young.

Size is a poor indication of species among North American swans, and all three varieties are glistening white in maturity. There are, however, differences of coloration as well as size and posture which distinguish the mute swan from the whistler or trumpeter. The mute swan is unique in the matter of neck curvature while swimming; the other two species hold their necks erect. The cygnets of all three species have dusky bills, but changes are noticeable by the time the birds are yearlings. The bill of the mute swan is then suffused with flesh color, and soon turns pinkish-red or orange, with a black tip and edges, and with a black fleshy knob at the forehead. The bill of the whistling swan turns entirely black, and that of the trumpeter becomes black with a narrow streak of salmon-red edging the mandibles. Both the whistler and the trumpeter have a small naked area between the eye and the base of the bill; on the whistler this area is often marked with a yellow spot (regardless of sex). The spot is lacking on all trumpeters, but since it is also lacking on some whistlers its absence is no certain indicator of species.

Differences in size are difficult to judge because all three species are of such magnificent proportions. With the neck held straight out as in flight, the whistling swan averages about forty-eight to fifty-five inches in length from tail to bill, and sometimes has a wingspread of over seven feet. The mute swan is slightly larger, and the trumpeter swan— often attaining a wingspread of more than eight feet and a weight of more than twenty pounds—is the largest of all North American wildfowl.

Voice, too, is an indicator of species. Swans are generally silent when on the water, but the mute swan is misnamed; though not vociferous, it will bark like a small dog to call its young, and it will hiss when angry, fanning out its wings in a threat posture. Some of the whistling swan's calls have the haunting quality of a flageolet or recorder or some other archaic flute, but the woodwind notes may be interspersed with a variety of long whoops, musical laughter, gooselike honks, cluckings, and murmurings. The trumpeter's call is, not surprisingly, a loud, low-pitched, extraordinarily clear and resonant bugling, sometimes followed by several higher notes. It has been likened to groaning, the keening of mourners, as if a dying race of creatures were uttering a lament. In 1795, describing wildlife observed during a journey from Hudson Bay to the Arctic, an early naturalist named Samuel Hearne compared the call of these birds to the sound of a French horn, but added that it was devoid of melody and often made him "sorry that it did not forebode their death." The words now seem a singular blend of petulance and prophecy.

There was a time when the Hudson's Bay Company—originally the Governor and Company of Adventurers into Hudson's Bay—took almost as great an interest in feathers as in furs. In justice, let it be granted that the role of the Company has not been as detrimental as that of the northern lumbermen or miners, and some of its field personnel have made significant contributions to the study of wildlife. Nonetheless, between 1853 and 1877 the firm handled 17,671 swan skins, most of them from trumpeters. Feminine taste at that time ran to plumage.

*Trumpeter's crude island nest in Alaskan marsh is safe from most predators.*

*Trumpeters sometimes permit intruders to approach closely, though swan pictured here is nervously alert.*

The number, as Peter Matthiessen pointed out in his study of American wildlife, was not staggering, but the trumpeter swan, like the whooping crane, evidently was never an abundant species, and the onslaught of the feather merchants was made deadlier by the loss of habitat. Trumpeter swans once bred from Alaska to Indiana. They wintered on all four flyways, with a sizable concentration appearing annually on Chesapeake Bay. Now the scant remnants of that population are confined to small areas in Wyoming, Montana, South Dakota, Idaho, Oregon, the Canadian Rockies, and Alaska.

The long white necks of swans often take on a rusty hue when the birds reside on certain waters. On the wet spruce muskeg of the Kenai Peninsula, the trumpeters stretch copper-stained necks for a better view of the latest intruders. Hungry Eskimos and Indians have been killing the nesting birds for centuries, but they are no longer the major threat. The muskeg is now seamed with roads, scarred by truck and caterpillar treads. Oil has lately been regarded as a more important resource in the Kenai than the swan, the moose, and the bear. The habitat continues to decline.

Fortunately, however, the protection afforded trumpeters for over half a century has been effective. Contrary to stories in the popular press, these swans are maintaining healthy numbers in Alaska and Canada, and are again on the increase in the contiguous states. During the early years of governmental conservation programs, field observers underestimated the population of trumpeters because they mistook the birds for whistlers. There is now reason for cautious optimism with regard to trumpeter swans.

The whistling swans, always more numerous, were also more reclusive and wary than trumpeters in the days before legal protection altered their habits. Moreover, their range was more extensive. Evidently they were never in danger of truly severe depletion, though their numbers were certainly diminished before the dawn of the twentieth century. Their recovery has been awesome; unprecedented numbers having been reported in recent winters on the Delmarva Peninsula. A photographer exploring a salt-water impoundment in Virginia in 1971 reported seeing "at least a thousand in one raft . . . and there were several more rafts in the distance."

Even during the days of unbridled shooting, swans were not killed in such great numbers as smaller wildfowl. The flesh of a single adult whistling swan, tough though it might be, could feed a number of people. Then, too, sportsmen tended to consider swans as spectacular trophies rather than game meat, and if a shooter carted dozens or scores of feathered carcasses from a marsh, their singular trophy quality was reduced by each addition to the mammoth bag.

Smaller birds such as redheads and other ducks were counted by pairs, and there is no doubt that the redhead population was endangered by gunners whose typical sunset boast was of "shooting fifty brace." On the other hand, at dusk of a February day in the 1870s, it was typical of a swan hunter on Currituck Sound to boast that in eight hours, "by dint of good fortune, I managed to bag nine magnificent swans." So wrote a correspondent for *Frank Leslie's Illustrated Newspaper,* reporting from Norfolk.

"Currituck is the home of the swan," he wrote, "the shores being beaded by the white bodies of the birds, the long, snowy line at a distance resembling foam. Several gun-clubs dot the low-lying sand-hills, whither the crack shots of New York and Boston repair during the season to commit fell havoc upon swan, goose and duck." However, he went on to say that, because of its distance from New York, the sound was "comparatively secure against the irruptions of Cockney sportsmen. None but the genuine lovers of the sport would travel so far 'for a blaze at the feathers.' "

His account describes the sound as being in some places eight miles wide, very shallow, and fringed with grasses sought by the wildfowl. The concentration of swans was very heavy in February, when new grass began to sprout. The sound has not changed as drastically as a host of less fortunate gunning localities. At that time, however, the writer described canvasbacks as being dropped in large numbers at Currituck and "disposed of on the waters to dealers at a dollar and ten cents the pair," and told how swans were "brought within range by 'blinds' and 'batteries,' the blind being a movable thicket, shifted and moored at will, within which lies concealed the boat of the wily hunter." Since he himself performed his eight-hour stint while reclining in a sinkbox "as though in one's coffin," it is reasonable to assume that he interrupted the "ordeal" with rest and dining periods, yet he probably could have killed many more than nine swans had he wished to.

The use of punt guns had been discontinued, he observed, because their loud reports frightened birds away from the sound, but he was equipped with two shoulder guns and found that both could be put to use during the gliding descent of the swans and before they could veer out of range. He made no distinction between trumpeters and whistlers, but from the evidence most of the birds must have been whistling swans and they were present in multitudes.

"The Sound was literally covered with feathered game. . . . Before us we could see hundreds of snow-white swans, some feeding, some arching their graceful necks over their wings, others with their wings set allowing the breeze to impel them gently along, and all enjoying themselves in the rays of a magnificent sun."

A guide had warned him to lead the birds well, for whistling swans can fly deceptively fast. They have been clocked at speeds of eighty miles an hour, but their massive bodies, long necks, and long, slow-beating triangular wings impart an appearance of regal leisure to a procession in flight. In the distance their heads may look like the points of slender lances, yet the overall effect is so ponderous as they approach that their speed is difficult to assess.

However, this is not to say that they are difficult to shoot down. Because of their size and momentum, they usually set their wings when they are still hundreds of yards from a swimming group they wish to join. They glide in toward the site where they will alight, circling at least once and on occasion several times as they descend lower and lower. Then, finally, they pitch in gracefully, usually without the splash of other birds and without extending their feet forward as do ducks and geese.

Pass shooting at speeding swans was rare. It was during their gentle, vulnerable descent that the guns boomed. There were plenty of shooting opportunities because whistling swans are both gregarious and restless, constantly moving about, leaving one group, joining another, seeking new areas of forage.

Whistlers have been beset by other hazards than the gun. During the long flights of fall and spring migrations they ascend so high that ice crystals form on their wings, sometimes disastrously. Once, in northern Pennsylvania during the spring of 1879, hundreds of them were forced down when their wings were laden with ice. Exhausted, they descended on pond and stream, field and village, and the helpless birds were clubbed by men and boys in a bloody orgy of meat hoarding. The birds have also been destroyed during northward flights by the "Niagara swan trap.'' In some years large numbers of them come to rest on the river above the great falls, and they are in danger of being carried over the cataract by the rapids. Though it probably does not happen with great frequency, Kortright reported that as many as two hundred have met death at one time when they were ''dashed against the rocks or crushed by the ice.''

Yet in spite of the perils, they have been manifesting a steady resurgence of numbers. They breed along the remote upper fringe of the continent, from the western and northern coasts of Alaska eastward across the top of the Yukon and the Northwest Territories, on the Arctic islands, and along the northwestern shores of Hudson Bay. Some of the westernmost breeding population follows the Pacific Flyway down into California and the Mexican peninsula of Baja California. But most of the birds fly southeastward across the continent to Chesapeake Bay and smaller waters in Virginia and North Carolina. Thus the Atlantic Flyway remains the winter quarters for many of the swans of North America, and in 1971 it was estimated that sixty-nine percent of the East Coast's swan population wintered in Maryland.

San Francisco Bay and its environs shelter almost as many swans, and odd fluctuations occur in some years. One winter brought more of the birds to the Pacific shores than to the Atlantic, causing a governmental research scientist to speculate that perhaps some swans change flyways from year to year.

The young are able to leave the nest by early July and to fly by mid-September. A month later they begin their autumnal journey. Observers have reported that ''as many as five hundred may form a single line, flying silently just over the shore line at a height of less than six hundred feet.'' As migration progresses, they tend to make longer, higher flights, and the attenuated lines break into small, probably familial chevrons. Wedge after wedge appears out of the sky at resting and feeding sites and at the winter destinations, where great rafts of birds are mirrored in the waters as they drift or swim, occasionally dipping their long necks to feed.

The mainstays of their diet today are wild celery, widgeongrass, and thin-shelled molluscs. Among the forces currently advocating a resumption of whistler shooting are representatives of the soft-shelled clam industry, and a relatively small number of sportsmen who feel that the swans not only reduce the food supply for game species but entice ducks away from their decoys. It is unlikely that swans pose a significant threat to the clam diggers and duck hunters. Only men of limited vision will compete for exclusive reign over constantly shrinking habitat that must be maintained and shared by all. However, there are bound to be a few Chesapeake gunners who cannot grin at the irony of swans unwittingly serving as live confidence decoys to lure ducks out of range where once huge artificial swans served as confidence decoys to lure them into range.

Though great numbers of swans find plentiful food and shelter at such sanctuaries as Eastern Neck Island and the Blackwater Refuge, even the very air is a contested region of habitat, principally because whistlers fly so high during migration. In 1962 a flock of swans struck a Viscount airliner flying at six thousand feet over Maryland, causing a fatal crash. Fatal collisions are infrequent, but the Federal Aviation Administration considers the birds an air hazard and has been engaged since 1967 in a long-term study of migration patterns and altitudes in order to prevent future accidents.

To facilitate the research, many swans must be captured. Some can be caught with nets (so incautious have they become during the years of protection); others are picked up after they have been tranquilized by doctored feed—an instance of wildfowl baiting for the benefit of both bird and man. Plastic collars of various hues are then affixed to their necks to designate the site of capture. Bird watchers, refuge personnel, and outdoorsmen are requested to report sightings of these birds as an aid in charting their movements. Small radio transmitters are also harnessed to some of the swans in order to track their flights. The signals can be recorded by receivers on the ground or in aircraft.

In one important experiment, a small plane has been used to follow a migrating flock, monitoring the radio signals. This has proved to be a difficult task because a small aircraft is likely to require refueling before the swans reach a resting site. Such mighty, perfectly adapted birds do not tire easily.

One of the scientists participating in this study is Dr. William J. L. Sladen, who is also engaged in swan research jointly sponsored by Johns Hopkins University, the Canadian Wildlife Service, the United States Fish and Wildlife Service, the United States Air Force, and state and provincial game departments. Dr. Sladen has noted that the spring migration of swans is accomplished in nonstop flights of two hundred and fifty to seven hundred miles, interspersed with feeding respites of ten to twenty-five days. During the long flights, whistlers can fly as fast as eighty miles an hour (though fifty is more typical) and at least as high as eight thousand feet. Sladen and others believe they can reach altitudes of ten thousand feet.

Since the flocks cross major airline routes, the importance of the research is obvious, and there is yet another reason for such study. "We also hope to use the swan as an indicator of man's effect on the environment," Dr. Sladen has remarked. "It is a conspicuous bird which needs resting and feeding places along its migration routes. Pollution may affect these birds more than other waterfowl, so the swans could prove to be a good indicator of environmental contamination."

In view of this need for environmental study, and of the relative tameness that has developed in this denizen of the Atlantic Flyway— and in spite of the swan's paltry competition for habitat coveted by clam mongers—it would seem advisable to continue the policy of protection or to relax it only slightly. A very limited season might be opened in some overpopulated areas, perhaps on a permit basis with a seasonal limit of one or two birds per hunter. There is no reason, after all, to regard the swan as a lesser trophy than the wild turkey.

Apart from uncharted migratory paths and the questions regarding the interaction of whistlers with the environment, there remains one more mystery to be solved in the study of the swan. It is the controversy, many centuries old, surrounding the legendary swan song which is said to be the call of death. Most naturalists have long agreed that the swan's dying song exists only in the fancy of poets. Frank Leslie's reporter on Currituck Sound mentioned a throat ailment "somewhat resembling diphtheria" which seemed to be fairly common in the 1870s and was offered as an explanation of the legend: an afflicted bird, no longer able to eat, "droops and dies, uttering a low sound which may probably have given birth to the poetical idea that swans sing before they die."

In 1887 the celebrated ornithologist Elliott Coues declared in his *Key to North American Birds* that swans lack the vocal apparatus necessary to produce the modulated death notes attributed to them. Yet, the no less eminent naturalist D. G. Elliot wrote in 1898 that he had personally shot a splendid swan at his club's gunning site on Currituck Sound and he had heard the dying song:

I am perfectly familiar with every note a swan is accustomed to utter, but never before nor since have I heard any like those. . . . Most plaintive in character and musical in tone, it sounded at times like the soft running of the notes in an octave. . . . We stood astonished and could only exclaim, "We have heard the song of the dying swan."

Those who have hunted much know that most animals succumb to death in silence, but sometimes a stricken creature will cry out. If those cries are a meager foundation for the fable of the swan song, an

*Trumpeter swans, America's largest wildfowl, cruise proudly
over their waters with necks held straight as schooner masts.*

anthropomorphic or supernatural aura is perhaps man's homage to the swan's magnificence. Several centuries before the birth of Christ, philosophers were seeking the hidden meaning of the way in which wild creatures die. In one of Plato's most famous *Dialogues,* a philosopher named Simmias expresses doubt that a man can anticipate death with equanimity. Socrates, who will that evening accept his cup of hemlock, replies to him in words that might be pondered by any man who is reminded of his own mortality upon touching the down of vanquished game:

Will you not allow that I have as much of the spirit of prophecy in me as the swans? For they, when they perceive they must die, having sung all their life long, do then sing more lustily than ever, rejoicing in the thought that they are about to go away to the god whose ministers they are. But men, because they are themselves afraid of death, slanderously affirm of the swans that they sing a lament at the last, not considering that no bird sings when cold, or hungry, or in pain. . . . The swans have the gift of prophecy, and anticipate the good things of another world. . . . And I, too, believing myself to be the consecrated servant of the same God, and fellow-servant of the swans . . . would not go out of life less merrily.

# Chapter Eleven

# GLIMMERINGS OF THE GOLDEN AGE

The golden age of American wildfowling reached its zenith on the Atlantic Flyway and was as real as the flyway itself, yet the era defies succinct definition or precise dating. It was a protracted period when shooting for the sake of sport rather than subsistence or profit had become not merely respectable but an adjunct of gentlemanly status. When prosperity bred a new hunger for recreation and when true affluence spawned luxurious gunning clubs. When the plain garb and guns of pioneers had given way among sportsmen to the English mode of dress, the English reverence for fine guns. When the beginnings of the Industrial Revolution and advances in agricultural technology had brought leisure for hunting, but also when those same developments were unwittingly allowed to undermine the environment of hunter and hunted alike. When marshes, prairies, and forests were still too vast for comprehension.

The last frontiers had not been paved or even glimpsed in the time of the nation's youthful exuberance. But because it was not a time without conscience, it was also marked by contradiction and paradox. Game appeared so eternally plentiful that market shooting was tolerated, even encouraged, yet a significant number of sportsmen expressed fears about the future of wildlife. The dwindling of game was permitted to continue, yet the very notion of sportsmanship was being linked to the novel concept of conservation. What had seemed Victorian elegance began to look like Victorian ostentation, vulgar gluttony, accelerating recognition of the virtue of restraint.

It can be argued, of course, that the origin of the golden age was pre-Victorian. Perhaps it began, in Maryland at least, with the granting of a provincial charter to George Calvert, the First Baron Baltimore. The Calvert dynasty immediately extolled the pleasure as well as the profit to be derived from hunting the profusion of swans, ducks, geese, and other game. In New Jersey perhaps it began in 1813, when one of the new republic's first hunting and fishing clubs was incorporated under state law. Operating under the impressive title of "The Fowling and Fishing Association of Upper Township, Cape May County," the club

*Illustration of "Woodcock Fire-Hunting" accompanied 1841 account of southern bird shooting by Thomas Bangs Thorpe. Article appeared in one of New York's sporting journals.*

was organized by men who wished to protect their outdoor enjoyment by protecting the habitat. The conviviality of such clubs was sometimes overshadowed by a reputation for game slaughter, but it is a fact that many of them soon imposed bag limits on their members.

On North Carolina's Currituck Sound the golden age may well have begun with the closing of the inlet below Corolla in 1824. After that, the sound's fresh water attracted unprecedented numbers of ducks—as well as hunters and, later, sinkboxes.

In Louisiana and other parts of the Deep South where the Atlantic and Mississippi flyways opened into vast wintering grounds, the heedlessness of the early golden age was in ample evidence by 1841, when the humorist Thomas Bangs Thorpe published an account of "Woodcock Fire-Hunting" in *Spirit of the Times.* In a scholarly anthology entitled *Hunting in the Old South,* Clarence Gohdes characterizes the *Spirit* as "a prominent New York journal devoted to the turf, sports, and the theatre." By 1841 New Yorkers were renowned for their love of the hunt. They were well acquainted with jump-shooting for wildfowl and with firelighting for both deer and ducks, but their manner of hunting woodcock was the conventional technique with pointing dogs. It may be presumed that they were delighted to read how, in two hours of "sport," Thorpe and a companion had killed nearly thirty birds— mostly woodcock but including a couple of quail and a meadowlark. "With old hunters," he added, "the average is always more, and a whole night's labor, if successful, is often rewarded with a round hundred."

Such hunting was best, he said, during late December, January, and early February, when the birds had all arrived from "as far north as the river St. Lawrence." The technique was to trudge through swampy woods at night, in the company of a Negro servant who carried a long torch consisting of burning pine knots in a metal basket attached to a pole. The recommended gun was "a short double-barrelled fowling-piece of small bore." No dog was needed, since "the birds show plainly from three to ten paces all around." Even in the dark, the shooting was

*Charles E. Whitehead's 1860 report of
duck shooting in Florida included scenes—
typified by woodcut reproduced here—
in remote wetlands that no longer exist.*

so easy that Thorpe's tone became somewhat apologetic as he wrote
that a good marksman "has only to overcome his astonishment, and we
will add, horror, at the mode in which he sees his favorite game killed,
to be a perfect master of woodcock fire-hunting."

His mention of horror is revealing: even then, sportsmanship was
gaining strength, and butchery could arouse indignation.

Although Florida was sparsely settled until long after the Civil War,
the golden age had dawned there, too, by 1860, as indicated in another
*Spirit of the Times* article. The writer, Charles E. Whitehead, re-
counted an organized duck hunt in the "Drowned Lands" of the Everglades. It seems almost beyond credibility that Whitehead's Drowned
Lands have been thoroughly parched in little more than a century. The
area where he hunted has been subjected to canal-building, channelization, and, worst of all, extensive drainage for the sake of sprawling
town developments, concrete swimming pools, and high-yield agriculture. In some places, salt water has seeped inland to devastate fragile
fresh-water ecosystems, and great stretches that Whitehead negotiated
by boat are now so dry that forest fires are a constant problem.

But in Whitehead's day, sportsmen from the northern cities made
yearly junkets to Floridian plantations where skiffs were used to sneak
up and flush marsh hens, teal, black ducks, mallards, swans, geese, ibis,
and other species. Whitehead wrote of paddling with his Negro guide to
the edge of a large pond, the further end of which was "literally
covered with ducks." He identified them as a huge flock of teal that had
flown over at dawn. "It was a beautiful thing to see so much graceful
life so bountifully supplied and protected by its own instincts in the
solitude. . . . Three or four arose first, and then the whole mass cleared
the water; and the beating of their wings on the surface was like a long
continued roll of thunder. . . . It seemed as if they could be counted by
tens of thousands."

Using a gun borrowed from him, Whitehead's guide brought a dozen
ducks down by discharging both barrels; then Whitehead himself fired
both barrels of his gun, "bored a hole through the black mass, and
twenty-seven ducks fell." Some years later, when the account was re-

*Frederic Remington, best known for his western
paintings, canoed on New York waters
and sketched duck-hunting scenes in Florida.*

printed in a book, a British reviewer commented that Whitehead's "shoulder-gun brings down an almost incredible number of ducks at a shot," yet it is not difficult to believe such feats when one visualizes tightly massed flocks rising from the water. Whitehead did admit to an occasional mediocre shot, though he followed the confession with lavish praise for his "little Mullen gun" which seldom missed. "Canny John Mullen," he rhapsodized, "in his grim little smithy in Ann street, working away with his brogue and his cunning hand—how often have I thanked him when he never knew it!"

Within the next two decades, baronial wildfowling clubs began to dot the flyway, and at the height of each season market gunners spent much more time guiding wealthy sportsmen than shooting birds for sale. It was not long before the American plutocracy—such men as J. Pierpont Morgan and George M. Pullman—began to visit the fashionable lodges, as did several presidents and a number of colorful celebrities. Frederic Remington, fleeing New York or returning from a western painting trip, shot geese on the Tidewater as well as the Mississippi; and traveling deeper into the South, he sampled guided Floridian ducking.

Bennett A. Keen, whose brother John had guided Grover Cleveland and J. Pierpont Morgan on the Susquehanna Flats, said that the kill from a sinkbox often numbered well over a hundred ducks in a morning,

*Sporting prints of 1800s
reflected comfortable notion
that forests and streams
would stay eternally
bucolic and game-rich.*

155

*"Mallard Hunting," 1888 painting by Sherman Foote Denton, shows trees in full foliage and men in light clothing—perhaps in spring.*

and the *Baltimore Sun* reported an estimated five thousand birds killed on opening day, November 1, 1893, at Havre de Grace. Shooting heroes —men whose ability with a gun seemed supernatural—became nationally famous during that era. Among the best-known was Jesse Poplar, an Eastern Shoreman whose multiple-kill ability from a sinkbox was legendary even in the days before the advent of repeaters. His pick-up man, Ed Cloak, claimed that "when old Jess shot, something was bound to fall." The claim gained support from the *Baltimore Sun* dispatch, which stated that "one of the most remarkable shots of the day was made by Jesse Poplar just before the wind sprung up. Seven coots sailed over his decoys. He killed them all, using two guns and firing both barrels." With two hundred and thirty-five ducks, he took the high score on that opening day.

But it was not necessary to be a superlative marksman during the golden age. There is no evidence that the average gunner then was more adept than today's typical wingshooter. In fact, he was probably less skillful, since optometry was hardly more sophisticated than the manufacture of guns and ammunition, and clay-target shooting was not yet so widespread a pastime as to furnish a gunning education for a large proportion of sportsmen. Strangely, the few men who did become shooting heroes met with disapproval from some of the more nearly mortal sportsmen, including Grover Cleveland.

It was not a matter of envy. Cleveland and some of his contemporaries simply deplored the practice of killing large numbers of ducks for no better reason than to demonstrate one's marksmanship. The conservation movement had begun to gain momentum, and it was given sudden impetus by the advocacy of such famous men as Cleveland and Theodore Roosevelt.

# AUTUMNS
# OF THE
# PAST

Mid-19th-century painting by H. Hill, entitled "Day's End," portrays mallard hunters in region of Susquehanna Flats. Like Remington's goose-hunting scene on preceding page, it exudes atmosphere of bygone seasons when flights of Canadas were continually crossing overhead, and ducks like gadwall (lower left) and eider were legal game even in spring and summer.

Having transformed hunting from
food-gathering necessity to
gentlemanly diversion,
Victorian sportsmen emulated
British customs and dress.
English elegance even permeated
advertisements for percussion caps.
Extravagant attitudes exhibited
ignorance that seemingly
limitless swamps and marshes
were fast dwindling.

Wildfowl was depleted by 1930's, when Great Duck
Depression struck pintails (left), European widgeon,
swans, and Canada geese (below) as well as most
other species. Reduced habitat has impeded full recovery,
and conservation battle must continue.

*Print shows Niagara Falls in 1831, when region teemed with game; man, boy, and dog were sure to return with generous mixed bag. Canvasback and redhead decoys recall later epoch, when carving of working blocks had evolved into fine art.*

Among American statesmen, Roosevelt has always been regarded as foremost in his devotion to the outdoors. Actually, although he was foremost in his contributions to conservation, he was not the ranking statesman of sport except with regard to the pursuit of big game. In both fishing and wingshooting, he was overshadowed by the massive figure of Grover Cleveland. The late Allan Nevins, a most meticulous historian, made a strong biographical case for Cleveland as an under-rated president and the kind of always-needed American idealist who "values character above everything else." As a sportsman, too, and a chronicler of sport, Cleveland deserves greater recognition than has been granted him. He was the personification of the golden age, or of its finer aspects at any rate, and an able spokesman for the ethic and esthetics of wildfowling sport.

Upon returning to his Princeton home after his second term as president, he wrote a number of sensitive, witty articles on hunting and fishing for popular periodicals. In 1906, almost a decade after he had left office and two years before his death, ten of these essays were gathered into a little volume by the Outing Publishing Company under the title of *Fishing and Shooting Sketches*. The book, with Henry S.

*At Centennial Exposition in Philadelphia,*
*rustic exhibits illuminated various aspects of outdoor life.*

The Mission of
Sport & Out Door
Life.

Watson's original pen drawings, was reprinted in 1966 by the Abercrombie & Fitch Library. Its renewed availability has gratified many bibliophilic sportsmen. One of its essays, "The Serene Duck Hunter," is justifiably described in the introduction as a gem of American sporting literature and is of particular interest in its revelations of the golden age.

Here and elsewhere, in the course of subtly deprecating his own shooting ability, Cleveland derided the "insatiable dead shot" both as a game hog and as a self-centered exhibitionist who failed to respect or appreciate the game. In an era supposedly characterized by game-hoggery, he condemned hunters "whose deadly aim affords them gratification only in so far as it is a prelude to duck mortality, and who are happy or discontented as their heap of dead is large or small." He also expressed disapproval of sporting magazines enlivened by "stories of inordinate killing," and to the modern reader he bequeathed a twinge of *déjà vu* with the regretful declaration that "our countrymen seem just now to be especially attracted by the recital of incidents that involve killing,—whether it be the killing of men or any other living thing."

His essay pointed out that the insatiable shooters—a minority among wildfowlers—gave "unintentional support to a popular belief, originating in the market shooter's operations, that duck shooting is a relentlessly bloody affair." And as to market shooting itself, he said, "Not a particle of sportsmanlike spirit enters into this pursuit, and the idea never enters their minds that a duck has any rights that a hunter is bound to respect. The killing they do amounts to bald assassination—to murder for the sake of money."

Grover Cleveland's authentic wildfowler in the twilight of the golden age was a member of the "Honorable Order of Serene Duck Hunters," a man hopelessly enamored of nature and sport, and hopelessly encumbered with foibles:

Therefore our serene duck hunter should go forth resolved to accomplish the best results within his reach, but doubly resolved that . . . he will betray no ignorance of any detail, and that he will fully avail himself of the rule . . . which permits him to claim that every duck at which his gun is fired is hit—except in rare cases of conceded missing, when an excuse should always be ready, absolutely precluding any suggestion of bad shooting. . . . When the ducks have ceased to fly for the day the serene duck hunter returns to camp in a tranquil, satisfied frame of mind. . . . He has several ducks actually in hand, and he has fully enjoyed the self-deception and pretense which have led him to the belief that he has shot well.

In another essay, "Summer Shooting," he lauded the gentle pleasures of shorebirding on Cape Cod while again decrying the gluttony that had passed for opulence until the flyway was endangered. After verbally placing a small-gauge gun "beside the rod and reel in making up a vacation outfit," he wrote:

In July or August the summer migration from their breeding places in the far North brings the shore-birds and plover—both old and full-grown young— along our Eastern coast. . . . Like other members of the present generation and later comers in a limited field, I have been obliged to hear with tiresome iteration the old, old story of gray-haired men who . . . on this identical ground, have slain these birds by the thousands. The embellishment of these tales by all the incidents that mark the progress of our people in game extermination I have accepted as furnishing an explanation of the meager success of many of my excursions; but at the same time my condemnation of the methods of the inconsiderate slaughterers who preceded me has led to a consoling consciousness of my own superior sporting virtues. . . . Any bag is large enough for me, providing I have lost no opportunities and have shot well. . . . I would not advise the summer vacationist who lacks the genuine sporting spirit to pursue the shore-bird. Those who do so should not disgrace themselves by killing the handsome little sandpipers or peeps too small to eat. It is better to go home with nothing killed than to feel the weight of a mean, unsportsmanlike act.

A third gunning essay in the collection concerned a duck-hunting trip to Virginia during which Cleveland and four companions occupied blinds (enhanced by tethered geese as well as wooden decoys, if one accepts the Watson drawings as part of the narrative) where they killed about a hundred and twenty-five ducks in four days. That would amount to seven or eight ducks per day for each hunter, not so excessive a number as one might expect from a party of northern swells during the unrestrained golden age. Apparently, one of Cleveland's motives in writing the account was to rebut a slander circulated by political enemies that his party had killed five hundred ducks on that trip. The historically significant insight is that a four-day kill of five hundred ducks by a quartet of sportsmen was regarded as shameful in 1906. Such pillage would not then have been a violation of law, as Cleveland candidly admitted, but—evidently in agreement with the majority of his wildfowling brotherhood—he was championing "sport versus slaughter."

It is also interesting to note that sportsmen, in contrast to market shooters, had come to regard as a point of honor their adherence to the few restrictive game laws that did pertain: "It is not a pleasant thing for one who prides himself on his strict obedience to game laws to be accused of violating these laws. . . . It is not true that I was once

AMERICAN SPORTING SCENE: SNIPE SHOOTING.

*Hunting was popular theme not only for lithographers but for
celebrated artists. Still-life is 1895 work by Richard LaBarre Goodwin.*

arrested in Virginia for violation of the game laws, or for shooting without a license."

Regarding the accusation of game-hoggery,

There was not one in the party who would not have been ashamed of any complicity in the killing of five hundred ducks, within the time spent . . . nor is there one of the party who does not believe that, if the extermination of wild ducks is to be prevented, and if our grandchildren are to know anything about duck shooting, except as a matter of historical reading, stringent and intelligent laws for the preservation of this game must be supplemented and aided by an aggressive sentiment firmly held among decent ducking sportsmen, making it disgraceful to kill ducks for the purpose of boasting of a big bag. . . . Those who hunt ducks with no better motives . . . merit the contempt of the present generation and the curses of generations yet to come.

If the limit he and his friends set for themselves seems high by today's standards, it should be recalled that the federal limit for many years was twenty-five a day, and Cleveland's rule was to shoot only as many birds as could be consumed by his family and friends—evidently a common rule even in 1906. In the same account he welcomed still other restrictions which, by implication, were no longer extremely controversial: "In Virginia they have a very good law prohibiting duck shooting on Wednesdays and Saturdays, and of course on Sundays. These are called rest days."

Two other small but interesting sidelights on the golden age are shed by this essay. First, Cleveland related a disagreement with his guide over the placement of wooden decoys "to the windward of us." Effec-

160

tive decoy positioning had long been understood by amateur shooters as well as the professionals, and Cleveland was mystified by the actions of a guide who may have been anticipating some productive market shooting after the departure of the supposedly ignorant sportsmen from the city. Second, he mentioned his accommodation at the "very comfortable club-house of the Back Bay Club, in Princess Anne County." It was a typical facility for the wealthy sportsman.

So far as the good things and the comforts of the club-house itself entered into our enjoyment of the trip, it would be strange if they did not present great allurement; for nothing in the way of snug shelter and good eating and drinking was lacking. It is not so easy, however, to reason out the duck hunter's eagerness to leave a warm bed, morning after morning, long before light, and go shivering out into the cold and darkness . . . to sit for hours waiting for the infrequent shot which reminds him that he is indulging in sport or healthful recreation.

Cleveland's conclusion was that "the duck hunter is born—not made."

Perhaps the last spokesman for the wildfowlers of that era was the late Nash Buckingham, whose hunting memories stemmed "from Christmas morning in 1888 in the huge, high-ceilinged room of our old home. Folks got up early to celebrate in those days. I recall my father standing near the grand piano that my mother played so beautifully and presenting my older brother and myself with a 16-gauge Parker hammer gun and a set of boxing gloves."

Though Buckingham gunned more often on the lower Mississippi than on the Atlantic Flyway, the reminiscences he published in the Winter, 1969, edition of *The American Sportsman* quarterly would light a gleam of recognition in the eyes of any sportsman in yesterday's South:

To observe today's remnant waterfowl populations in contrast to what I saw in 1890 is but to mourn. Land drainage was in the distant future. Lumbering of the South's great hardwood and cypress stands was in its infancy. Lakes and marshes . . . lay in pristine condition for wildlife.

I gunned most of the famous duck clubs and public shooting grounds of the time: Reelfoot Lake; Big Lake; on down the Mississippi along which Menasha, Lakeside, Mud Lake, Beaver Dam, Wapanoca, and a hundred others were located; and all the way to the famed Delta Duck Club in Louisiana. . . . You really saw ducks and geese then! But even so, after 1890 the swans disappeared. . . . I saw them vanish suddenly from a favorite hangout, the Wapanoca Outing Club, as if they had evaporated. . . .

I have in an old diary a record of a Christmas-week hunt at Wapanoca in 1898. The daily bag limit imposed by the club was fifty ducks and as many geese as one could lower. That day at Wapanoca there were twenty-two guns: sixteen members and six sons home from college. Everyone bagged his limit of ducks, and thirty-eight Canada geese were killed. Today on Wapanoca—now a federal wildlife refuge—you would be lucky to bag a federal limit of four ducks.

I have another record of Wapanoca, on Washington's Birthday in 1901. Judge Gilham and I bagged an even hundred pintail, mostly drakes, before ten A.M. Not like the ducks in today's skies.

It didn't take long for wildfowl marketeers and sportsmen to realize that they could sack more ducks, geese, and birds per day—and faster—with the comparatively lightweight, six-shot repeaters than with the ponderous, heavy-recoiling, near-cannons they had been using. . . . As early as 1906 the Boone and Crockett Club passed resolutions calling for action against the autoloaders. The first of the repeaters I saw and used were the 12- and 10-gauge lever-action weapons by Winchester. My 12-bore cost about $15. . . .

My earliest recollection of a pump gun was a Spencer patented in the Eighties. It threw its discharged shells almost straight up. . . . There was also a pump gun called the Burgess. . . . The repeating action was in the pistol grip. Instead of pumping back and forth with the left hand, you shoved the right hand and pistol grip forward and pulled it back to reload. . . . There also appeared a two-shot pump gun by the Young Arms Company of Springfield, Ohio, the invention of Charles "Sparrow" Young, destined to become a renowned hunter and trap-shooting champion. . . .

In 1892, Mr. Irby Bennett, Winchester's district manager in Memphis, announced that his company was coming out with a new 12-gauge, and that he would give the first one as a school prize for the best grades. My brother Miles won it, but gave it to me. It has been handed on to my nephew and is still in use. . . .

I recall a day at old Wapanoca when, after having bagged the federal limit of twenty-five ducks at the head of Long Pond, I was rowing my duckboat homeward through Little Lake. Scads of ducks were coming in. . . . We went wood duck hunting in those long-gone days from late summer until early fall, being on our stands well before daylight. With dawn came one of the most beautiful flights and sights ever witnessed by outdoorsmen. First came slowly drifting thousands of little white herons, followed by a myriad of blue herons. Immediately following came the great white herons—or cranes, as we knew them. Their flocks were separate and distinct. They were going somewhere, and would do the same tomorrow.

Then came the wood ducks. Leaving their roosting area in a portion of ponded cypress, they flew in small family groups and larger bunches, and decoyed nicely to wooden teal decoys. Bag limits were, as I mentioned, fifty birds per day.

The famous Hatchie Coon Club above Marked Tree, Arkansas, on the once-beautiful and unpolluted St. Francis River, is an example of what real sport could be in those days. No automobiles, no outboard motors, no deep-freezers, and no market hunting on wood ducks. . . . We would pole upstream to the saw-grassed spreads, shoot a limit of fifty woodies, then catch a limit of bass en route to the clubhouse. . . . These scenes were being enacted all over the South in the period 1889–1909.

Life at such a shooting and fishing club as old Hatchie Coon (three or four clubhouses have burned, but each has been replaced along the same plan) was

*Romantic outlook of golden age, carried into 20th century, glows in
sketch by Charles Dana Gibson, one of America's most famous illustrators.*

tranquility itself. . . . The clubhouse, built high on stilts above the crest line of
the inevitable floods, provided a huge living room, cavernous fireplace, dining
room, and kitchen. Sleeping quarters and baths accommodated goodly parties. . . .

Clubhouse keepers at Hatchie Coon offered wonderful seasonal cuisines. Pre-
breakfast coffee was slashed with choice brandy and a lump of butter; quite an
eye-opening beverage if you've a mind to try it. . . . I'd still like to meet up with
a big pan of Bryant Cole's shredded coot, potatoes, and tomatoes, sugared and
baked with a top dressing of crumbs and parmesan cheese.

However, by 1900, when I was twenty, I knew that change was on the way. I
had seen prairie chickens and swans vanish. . . . The outdoor magazines wrote
chilling reports on the dwindling wildfowl resources and the abuses of upland
game. I realized, too, that some of the clubs . . . were cutting their daily bag
limits from fifty to forty ducks. Yet market hunting continued, and commercial
interests mocked the early voices that were protesting the decimation of the
flocks and calling attention to the destructive droughts and drainage on the
breeding grounds in Canada. . . .

It was not until 1916, however, that President Wilson signed the Treaty between the United States and Great Britain for the Protection of Migratory Birds
in the United States and Canada [now known as the Convention, implemented by the Treaty Act of 1918]. Opponents of conservation fought on until
1918, when by decision of the United States Supreme Court the treaty and
statute were upheld. For the majority, Justice Oliver Wendell Holmes declared:
"But for the Treaty and the Statute, there might soon be no birds for any powers
to deal with. We see nothing in the Constitution that compels the government
to sit by while a food supply is being cut off and the protectors of our forests and
our crops are destroyed."

. . . Thus the beginning—some fifty years ago—of the endless struggle to keep
game birds in the sky.

# Chapter Twelve

# CONSERVATION UNLIMITED

William Penn and a few other farsighted colonists had expressed concern for the future of America's lands and game, but conservation laws remained as isolated—hence, as feeble—as the state legislatures or private groups supporting them until the nation began to recover from the devastation of the Civil War. If, for example, Rhode Island's moratorium in the 1840s on the spring shooting of wood ducks, black ducks, woodcock, and snipe was an admirable restraint, it was of small avail as long as spring seasons were tolerated everywhere else. Not until the 1918 passage of the Migratory Bird Treaty Act was there a nationwide prohibition of spring wildfowling.

Similarly, in 1852 New Jersey had forbidden the shooting of ducks and geese from any boat or floating device more than fifty feet from shore; in 1859 night hunting became illegal on Barnegat Bay and on the Manasquan River; and the state ended all night hunting as well as shooting from sail and steamboats in 1879. But those practices continued in many nearby localities.

Progress was predictably slow. The idea of nationally reserved wildernesses or parks had been advocated since the 1830s by celebrities as influential as George Catlin, Thoreau, Emerson, Lowell, Horace Greeley, Oliver Wendell Holmes, and many more. But Yellowstone National Park, established in 1872, was the first of its kind. Soon thereafter, pressure groups began to exert a stronger influence, partly because they were led by famous men, but perhaps to a greater degree because they were publicized in the pages of outdoor magazines, and because voters were becoming more sophisticated in the use of their power over legislators, elected officials, and the press. The first federal wildlife refuge, authorized by President Theodore Roosevelt, was established at Pelican Island, Florida, in 1903, under the joint sponsorship of the government's Biological Survey and two such pressure groups—the American Ornithologists' Union and the Audubon Society.

George Bird Grinnell, editor of *Forest and Stream,* had helped to organize the American Ornithologists' Union in 1883. He was instrumental in the formation of the Audubon Society a few years later

*Wildfowl need more refuges like this wintering haven of mallards.*

(though he afterward found himself at odds with this group), and with Theodore Roosevelt he founded the Boone and Crockett Club in 1887. Men like the painter Albert Bierstadt, Senator Henry Cabot Lodge, and Generals Sherman and Sheridan were quickly attracted to this club, which was largely responsible for the establishment of the National Zoological Park in Washington, D.C., and the New York Zoological Park, as well as the passage of the 1894 Park Protection Act which sought to prevent the commercial exploitation of federally owned parklands. Publications such as *American Field* had carried advertisements a few years previously that described birds and beasts as gifts of God intended solely for man's pleasure and profit; in 1882 *American Field* recommended a closed pigeon season from March through May.

In 1886 the Department of Agriculture was augmented by the Division of Economic Ornithology and Mammalogy. This federal agency for the study of birds and mammals soon became the Division of Biological Survey; it achieved Bureau status in 1905 (and after many years of effective service was merged in 1939 with the Bureau of Fisheries to form the United States Fish and Wildlife Service). During the same period the Smithsonian Institution, reorganized with a new emphasis on research by Louis Agassiz of Harvard in 1876, conducted outstanding studies of migratory birds. By demonstrating the need for conservation, the findings of these agencies encouraged reform.

The proponents of conservation were also fortunate in having the unreserved sympathies of three successive United States presidents— Cleveland, McKinley, and Roosevelt. The first two of these men set aside large forest reserves, and Theodore Roosevelt expanded the program. Roosevelt's Reclamation Service, headed by Frederick H. Newell, also began the project of irrigating vast arid regions. It must be admitted that in this process valuable habitat was too often ruined, but the Service later provided waterfowl refuges on some of its impoundments, setting a beneficial precedent. In addition, Roosevelt appointed Gifford Pinchot to head the Forest Service and then, in 1908, the National Conservation Committee. It was Pinchot who coined the term "conservation" in its modern sense.

He was not a preservationist but a utilitarian who believed in multiple use of the wilds. There is no question that he made mistakes, and some of his schemes—such as flooding the Hetch Hetchy Valley to provide water for San Francisco—enraged John Muir's opposing faction of more radical conservationists. Controversy still causes tension between neo-Muirites and those of the Pinchot persuasion. A fair appraisal must grant significant contributions by both men. Pinchot was the chief engineer of Roosevelt's conservation machinery, and by awakening public support for wise land use he prepared the way for major reforms soon to come. Muir, the botanist, wilderness wanderer, protector of Yosemite, was an equally persuasive publicist whose writings awakened a new, healthily proprietary interest in the outdoors. The Sierra Club, founded in 1892, has faithfully followed his precepts, just as the Wilderness Society, the Wildlife Management Institute, and other concerned groups have adhered to the practical game-management and wilderness-preserving policies propounded in the 1920s by such innovators as Aldo Leopold and Robert Marshall.

The early conservation movement was blessed with an astounding array of gifted writers. It was as if a new evangelism had inspired the country's finest and most determined minds. G. O. Shields, editor of *Recreation Magazine* and coiner of the term "game-hog," was among the most vociferous; and the romantic John Burroughs, still widely read, was among the most convincing. Burroughs employed an earthy, pragmatic approach that brought readers close to nature and that sometimes led to insights of the kind now being repopularized by modern ecologists. "Nature," he remarked, "does not care whether the hunter slay the beast or the beast the hunter. She will make good compost of them both, and her ends are prospered whichever succeeds."

There were also successfully militant scientists, such as the ornithologist Elliott Coues, author of *Key to North American Birds;* William T. Hornaday, author of *Our Vanishing Wildlife,* curator and taxidermist for the Smithsonian, and later director of the New York Zoological

Snowy egret was among
species saved from
ravenous plume marketers.

Park; and Henry Fairfield Osborn, president of the New York Zoological Society and writer of a bristling preface to Hornaday's book. Osborn declared in 1912:

Nowhere is Nature being destroyed so rapidly as in the United States. Air and water are polluted, rivers and streams serve as sewers and dumping grounds, forests are swept away and fishes are driven from the streams. Many birds are becoming extinct, and certain mammals are on the verge of extermination. Vulgar advertisements hide the beauty of the landscape, and in all that disfigures the wonderful heritage of the beauty of Nature to-day, we Americans are in the lead.

Fortunately, the tide of destruction is ebbing and the tide of conservation is coming in.

In the following year two farreaching measures were enacted for the conservation of wildfowl. The ravages of the feather merchants had been exposed by Audubon investigators and indignant writers, and New York's Bayne Bill had prohibited trade in wild-bird plumage in 1910. In 1913, the federal Tariff Act barred importation of such plumage. At the same time, the Weeks-McLean Act awarded responsibility for migratory birds to the Biological Survey. This meant that wildfowl came under federal jurisdiction. Test cases were filed to challenge the law, of course, but the obvious interstate nature of migration guaranteed a favorable outcome.

Earlier chapters of this book have explored the magnificent ramifications of the 1916 Convention for the Protection of Migratory Game Birds in the United States and Canada, as well as the implementation of the Migratory Bird Treaty two years later. The major points of these measures can be listed briefly: Endangered species of wildfowl and shorebirds gained full protection; spring shooting ceased; the federal government confirmed its right to impose bag limits; and marketing

*Shrinkage of public hunting areas was subject of cartoons as early as 1879.*

167

restrictions were made so stringent that commercial shooting expired.

If the government had been equally adamant and farsighted in dealing with pollution, urbanization, drainage, land speculation, and other destructive inroads into habitat, it is unlikely that the environment would be imperiled today.

After 1918, the procession of reforms was slower but remained rather surprisingly steady. In 1925 the badly needed Alaska Game Commission was set up, and in 1934 the Migratory Bird Hunting Stamp Act inaugurated the sale of "Duck Stamps"—in effect, federal hunting licenses for wildfowlers.

It was the time of the Great Duck Depression. Brant were being starved almost to the point of extermination by the blight of eelgrass, which was also engendering a famine for other species. "Reclamation" programs and independent private and cooperative projects had succeeded in draining new millions of wetland acres in the United States and Canada. Benighted agricultural practices had exacerbated the results of drainage, sowing drought and reaping the infamous Dust Bowl. Immense prairies were devastated.

If the famous conservationist-cartoonist Jay N. "Ding" Darling, then the newly appointed chief of the Biological Survey, had known that the decade would soon be marked by another calamity—a series of severe hurricanes that would flatten most of the standing dead trees in the northeastern United States and Canada—he might well have given up in despair. Those hurricanes were to remove countless nesting sites of wood ducks and golden-eyes, thereby reducing still further the wildfowl population. But before the storms struck, additional new measures had been taken for the propagation of migratory birds.

In 1934, President Franklin D. Roosevelt appointed Darling and Aldo Leopold to a committee on waterfowl restoration. Through the medium of his popular newspaper cartoons, Darling persuaded a large segment of the public that help for waterfowl would also be of benefit in ending fierce, prolonged drought conditions. It seems incredible that in the midst of a great financial depression, Americans were willing to spend money to cure a great waterfowl depression, but duck salvation was cleverly combined with relief programs. The Darling forces managed the reflooding of the Malheur Refuge in Oregon, an early step in proving the validity of restoration theories. Funds acquired through the sale of duck stamps would halt erosion and restore the water table, salvaging stricken lands and helping to revive the economy. Labor was provided by the recruits of the Civilian Conservation Corps and the

*Handsome sketch of mallards by Ding Darling became first duck stamp, followed by fine Frank W. Benson etching of canvasbacks.*

Works Progress Administration—two New Deal devices for the employment of the jobless.

The Stamp Act was passed less than two months after Darling became head of the Biological Survey, and he was told that a stamp design was needed immediately if the stamps were to be engraved, printed, and distributed to all post offices sufficiently in advance of the hunting season. Darling sketched some rough design ideas on half a dozen sheets of cardboard and turned them over to Colonel Hal Sheldon, chief of public relations for the Survey. A week passed and no one said anything more about his ideas, so he asked Sheldon if any of the sketches were deemed worthy of refinement. He was informed that the Bureau of Engraving had already selected one of the drawings and was at work on it. The design was a hasty sketch of two mallards dropping in. Darling was furious. He had envisaged a polished portrait of the kind that has become traditional on duck stamps. However, as the conservation writer John Madson has pointed out, that first stamp "had a rough strength that fit the new program well." Its hard-lined sweep and animal vitality were reminiscent of Dürer's woodcuts.

Since then, duck-stamp sales have financed incalculable improvements in habitat as well as wildfowl research and the maintenance of existing habitat. Many philatelists consider the series to be the world's most beautiful revenue stamps. This is hardly surprising since they have reproduced wildlife portraits by some of the country's best-known artists: Frank W. Benson, Richard E. Bishop, Roland Clark, Lynn Bogue Hunt, A. Lassell Ripley, Stanley Stearns, Edward J. Bierly, Edward A. Morris, and Maynard Reece, among others. The selection is an annual competition whose reward is honor (there is no financial recompense) and the satisfaction of helping to conserve wildfowl.

The cost of the first stamp was a dollar. Six hundred and thirty-five thousand were sold. Over the years the price has climbed to five dollars, with little complaint from hunters. Sales have risen until the annual average hovers around two million or so, thus bringing yearly revenues to about ten million dollars. As of 1971, the income from duck stamps totaled a hundred and seventeen million dollars. During the first thirty-five years of sales, the income purchased six hundred thousand acres of refuge lands for migratory waterfowl, plus easements prohibiting drainage of yet another six hundred thousand acres.

Until 1960, these funds paid for acquisition, maintenance, and administration; since then, the money has been used solely for the purchase of additional lands, with other revenues being used to meet related expenses. In addition, Congress has authorized the appropriation of one hundred and five million dollars to accelerate the acquisition of habitat during the fifteen-year period ending in 1976, and over fifty-three million dollars of this fund has been appropriated as these lines are written.

Indirectly, duck stamps have contributed more than their face value to the preservation of game-bird habitat. They have provided valuable publicity for the cause of conservation, and some of the rare older issues as well as complete sets have been resold by such organizations as Ducks Unlimited and the Izaac Walton League to stamp collectors, connoisseurs of wildlife art, and concerned sportsmen. The proceeds have, of course, been used for the increase or maintenance of habitat.

In 1933, the last thirty-three trumpeter swans were counted at Red Rock Lake in Montana. Four years later, the Red Rock Lake National Wildlife Refuge was established. Chiefly funded by duck stamps, it has been credited with saving the trumpeter from extinction. Several hundred of the swans nest there today, and additional colonies are breeding at refuges as widely separated as Oregon and South Dakota. Duck-stamp income also established and for many years maintained the Lake Agassiz Refuge in Minnesota, where the land had been drained early in this century, only to prove agriculturally useless. The region's few farmers were forced out by the drought of the 1930s. But like many areas left barren by misguided land management, the acreage has been restored to environmental usefulness.

While assuring that many game species will survive and prosper, such refuges also provide hunting grounds for sportsmen, often in regions where private land ownership and industrialization have severely decreased all other natural habitat for *homo sapiens*. Public hunting is permitted on considerable portions of some of the larger waterfowl refuges such as North Carolina's Mattamuskeet, one of the most renowned of the Atlantic Flyway's goose-shooting sites.

The idea of combining hunting opportunity with refuge maintenance had been promulgated as early as 1922 by Dr. Edward W. Nelson of the Biological Survey and Ray P. Holland of *Field & Stream* and the American Game Protective Association. Their proposed "Public Shooting Grounds Bill" provided for the acquisition of refuges, portions of which would be open to hunters who would pay fees for shooting privileges and thereby help support the refuges. Strong opposition was led by Dr. Hornaday and was initially supported by *Outdoor Life,* the infant Izaac Walton League, the Camp-Fire Club, Aldo Leopold, Fiorello La Guardia, and others who feared that the refuges would turn into federally maintained shooting preserves which would produce no surplus of birds. Advocates of the Nelson-Holland proposal included the Boone and Crockett Club and the National Association of Audubon Societies. Dr. Nelson maintained that because of food scarcity on the wintering grounds, a great increase in the wildfowl population was not then desirable, and that the birds would benefit most from state-regulated shooting combined with federally regulated game management on refuges. He was seconded in an Audubon bulletin by Charles Sheldon, who was also an influential member of the Boone and Crockett Club. Though the bill was defeated, the Camp-Fire Club, the Izaac Walton League, and *Outdoor Life* eventually saw the light of reason—in the form of sport-supported refuges which eventually materialized in the acquisition program authorized by the 1929 Migratory Bird Conservation Act.

Their judgment was vindicated in the 1930s, after the overly liberal federal bag limits were reduced and new refuges were financed by sportsmen. Furthermore, Dr. Nelson's theory was proved delightfully incorrect because the refuges did greatly increase the wildfowl population, while also correcting the food scarcity by providing new forage on the wintering grounds.

In 1936, two years after the passage of the Duck Stamp Act, President Roosevelt called the First North American Wildlife Conference,

*Maps show National Wildlife Refuges in Northeast and Southeast; acquisition program has been in progress since 1929.*

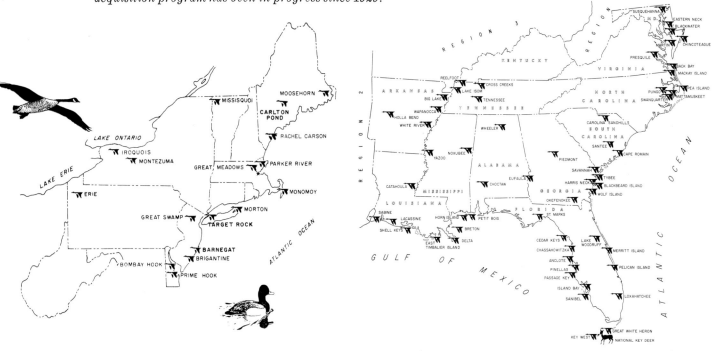

*Large proportion of refuge acreage in growing national system is devoted to migratory birds, but gains do not yet outstrip losses.*

conducted (as it has been in succeeding years) by the Wildlife Management Institute. Also in 1936, the United States and Mexico concluded a treaty modeled on the 1918 Canadian-American treaty and thus extended governmental control of migratory birds over virtually the entire continent, from end to end of all four flyways.

The treaty was ratified the following year, when yet another victory was achieved for hunter and hunted alike. This was the Pittman-Robertson Federal Aid in Wildlife Restoration Act, one of those rare instances in world history when a tax was levied at the behest of those to be taxed. Unlike earlier refuge proposals, the Wildlife Restoration Act aroused little controversy among the nation's hunters. They volunteered to be tithed in order to assure the survival of game, and their wish was granted by the imposition of an excise tax on firearms and ammunition. This tax now provides about thirty-seven million dollars per year in Pittman-Robertson funds for the benefit of wildlife. Peter Matthiessen's *Wildlife in America,* a scholarly work which is not very kindly disposed toward hunters, concedes that "extensive programs of state and university studies and resource management" are supported by these funds.

There is no question that sportsmen have contributed more than any other segment of the population to governmental conservation programs and thus to the restoration of the environment. In less than fifty years, hunters alone have willingly paid an estimated 2.2 billion dollars for these programs, and the figure includes no donations or membership dues to conservation organizations but only state and federal fees: 1.6 billion dollars for state licenses, 117 million for duck stamps, 438 million in excise taxes. It is heartening to realize that all three of those figures will have continued to rise during the brief period between the writing of this book and its publication.

In the years since the enactment of the Pittman-Robertson Act, further advances have been made. Shortly before World War II, the Biological Survey and the Bureau of Fisheries were merged into the Fish and Wildlife Service of the Department of the Interior, thus removing their administration from the commercial temptations of the Departments of Agriculture and Commerce. (The latter agency now has charge of commercial fisheries only.) And in 1956 the Water Pollution Control Act was passed, providing federal financial aid for communities willing to improve sewage treatment and begin to face the problems of industrial pollution. There has been a law ever since 1899 requiring industrial polluters to obtain a permit before dumping foreign matter into navigable waters, but the weaknesses of such laws—even when not visible along a shoreline—have been made obvious by a new generation of conservation writers. Some of the traditions of Muir, Pinchot, and Burroughs have been sustained in more recent decades by such informed writers as Paul Errington, Loren Eiseley, and Rachel Carson. Though legislators have failed to keep pace with the despoilers of habitat, there has been a new surge of concern during the last decade, and many are the congressional proposals even now being considered.

Much has been said regarding the government's role in saving the flyways from the annihilation of habitat that sometimes passes for technological progress, and reference has also been made to several private conservation groups. One of the latter merits a separate accolade. In 1968, the Post Office Department commemorated the fiftieth anniversary of the Migratory Bird Treaty by issuing a six-cent postage stamp picturing two wood ducks in flight and bearing the legend "Waterfowl Conservation"; one of the press releases distributed by the postal authorities in connection with this special issue graciously de-

*Multitudes of snow geese lift off foraging area in federal refuge.*

scribed the stamp as a tribute not only to governmental conservation efforts but to those of Ducks Unlimited, a sportsmen's organization which has done more than any other group to maintain and increase migratory birds.

It is fitting that the incalculable benefits of Ducks Unlimited had their origin on the Atlantic Flyway. In 1930 a New York foundation, known as More Game Birds in America, Inc., undertook a five-year survey of duck populations, together with investigations into the causes of decline. Though predatory birds and animals were taking a heavy toll (they had been less severely affected than herbivores by man's proximity), drought was found to be the major factor. Restoration and management programs were inaugurated on the American and Canadian breeding grounds, but in that period of economic calamity it was impossible to obtain sufficient governmental grants. The leaders of the foundation, agreeing that the burden would have to be borne voluntarily by sportsmen, sponsored an association to finance and conduct the work. In subsequent discussions, the leaders of this private organization settled on the name Ducks Unlimited, and DU—as it has been called by legions of dedicated members—was incorporated in the District of Columbia on January 29, 1937.

Since that date, it has become International Ducks Unlimited, with closely affiliated Canadian, American, and Mexican components, and with a membership spanning the continent.

Members have contributed astounding sums (as well as their time and skills) and notable philanthropists have enlisted under the DU banner. As a single typical example, the Western Pennsylvania Project of the 1960s was made possible by donations from Seward P. Mellon and Richard P. Mellon as a memorial to their parents, General and Mrs. Richard K. Mellon. This project of the western-Pennsylvania members of DU consists of 1,279 acres of water, and twenty-four miles of shore-line north of Meadow Lake, Saskatchewan. Before water-control work and diking was completed, the wetlands there were subject to drought in some years and to disastrous flooding almost every year when the Beaver River was swelled by the spring runoff. Now the marshes are blessed by an ideal water table and an abundance of coontail, bulrush, water milfoil, sage, flagreed, flatstem pondweed, and other vegetation needed by nesting waterfowl. The principal nesting species include mallard, pintail, blue-winged and green-winged teal, gadwall, widgeon, canvasbacks, redheads, and bluebills. The Pennsylvanians and others on the Atlantic Flyway enjoy part of the bounty, and goodly numbers of the ducks are found on the other migratory corridors as well.

By 1971, DU had collected and used some twenty-four million dollars for the acquisition, restoration, improvement, and maintenance of productive wetlands. It had spread its protective wing over two million acres of water—a figure which, though it might boggle the imagination of the city-bound, does not include the rich and enormous fringe habitat bordering the water. The funding of Ducks Unlimited has steadily increased every year for the last decade, with each season setting a new record and annual contributions surpassing two million dollars for the past few years. In 1971, over three million dollars was collected.

Perhaps the rising support during the last decade was a response to a crisis. In 1956, an estimated hundred and thirty million ducks had traversed the American flyways. It was a superlative year. The autumn of 1962 was the most miserable in recent history, with the estimate plummeting to seventy million. The decline was caused by another of the extended and devastating droughts that have scarred the century.

*Canadas set down on grazing field at Blackwater Refuge in Maryland.*

The six or eight million potholes dotting the prairie provinces in a wet year had withered to as few as six hundred thousand. Ducks Unlimited remedied this ecological dry rot by spreading and managing the northern marshes, and the decade ended with banner crops of wildfowl.

There are shooters who can recall not one but four Great Duck Depressions. The first became critical in 1915, the year before the United States and Canada proclaimed a new era of international conservation. The second occurred in the 1930s, when Mexico joined the waterfowl alliance and when the birth of Ducks Unlimited accompanied the passage of some of America's most important wildfowling legislation. The third occurred immediately after World War II when bag limits and seasons were severely curtailed and invaluable migratory studies were carried out. The fourth occurred in the 1960s, a decade marked by an unprecedented clamor for legislation to alleviate pollution and halt the destruction of habitat. It was also a decade marked by the emergence of a new breed of sometimes misguided, sometimes fanatic, but totally committed environmentalists. Though environmentalism may be only faddish, it is possible that America has dragged itself to the brink of an ecological revolution that presents an alternative to self-destruction. For calamity is the mother of reform.

*Saskatchewan hunters prepare to leave one of Canada's immense fields,*
*which support geese and other birds that winter on every flyway.*

# Chapter Thirteen

# PROSPECTS

There have been warnings that unless the proliferation of *homo sapiens* is halted within the next generation or two, the planet will not support the human population. Why, then, concern ourselves with the capacity of one flyway or of the world to support ducks? A straw-grasping answer is that wildlife may help to support humans, at least for a small while, after most of our presently utilized resources have approached exhaustion. That answer is only a plea for a brief stay of execution. A response more to the point is that each resource we learn to conserve teaches us how to conserve other resources, thereby pointing the way toward the world's ultimate salvation. We have at last accepted the obvious presumption that nature's balance must be restored if we are to survive. Another response is equally valid, for the word "salvation" is used here in more than its physical sense. The-quality-of-life has become a catch phrase as a result of its own attrition. What is left must be conserved, and some of what is gone must be regained, because the instinct of self-preservation is not a wish to survive alone, on a desolate planet shrouded by a barren sky.

Some further deterioration is probably inevitable before we reach a turning point, but our very desperation is cause for optimism since it goads us to seek remedies. In 1965 there were fewer than three hundred officially designated National Wildlife Refuges in the United States. Now there are about three hundred and thirty, with more being acquired at an average rate of three per year. They cover nearly thirty million acres, some forty-seven thousand square miles, without including more than a million additional acres of wetlands held through purchase and easement.

On the heavily urbanized Eastern Seaboard there is no plethora of appropriate rural sites for refuges, but rural seclusion is not always essential. New York's Jamaica Bay is a city-operated sanctuary created in the 1950s by Parks Commissioner Robert Moses. It has survived next to the Kennedy International Airport, an absurd location far better than none at all. Its existence is rendered precarious by dumping, by air

*Huge flock of Canadas, flushed from eastern cornfield,*
*is far more common sight today than two or three decades ago.*

and water pollution, by periodic threats of airport expansion, and by proposals for "better" uses, but the Bay may yet be preserved as part of a national recreation area.

Excluding sites more properly assigned to the Mississippi Flyway or where two migrational corridors blend, the Atlantic Flyway has more than sixty National Wildlife Refuges, close to twenty percent of the total. A large proportion of refuge acreage is devoted to migratory birds (and particularly waterfowl). If ideal conditions prevailed eternally and no other factors were involved, the present wetlands would be more than adequate. But an estimate published by the National Wildlife Federation shows that "a million and a half acres a year of prime wildlife habitat are being taken for highways, airports, housing and industry. Government at all levels is attempting to set aside 'green acres' and wildlife living space, but acres lost still outstrip those that are saved." Though the obliteration of habitat was reduced to about a million acres in 1970 and again in 1971, losses continued to outstrip gains. New refuges are acquired annually, but a few are also lost. Federal maintenance of four ceased in 1970; some of those wetlands will continue to support birds, but not as well as previously. In the same year, the government designated nearly three million refuge acres for possible future disposal. So far, little of that land has been relinquished, but its eventual fate will depend on public pressure—a rarely predictable factor.

It was also in 1970 that the Portland, Maine, *Sunday Telegram* reported a program of the state highway commission for controlling unwanted vegetation: "Against cattails they use Radapon, a herbicide which is harmless to broadleafed plants." That the cattails may, in spots, have created some genuine problems is irrelevant to the attitude expressed in a brief editorial aside: "Cattails, if not controlled, can soon clog drainage ditches and create a swamp."

Three years previously, Anthony S. Taormina of the New York State Conservation Department expressed concern about the tendency for wetlands "to be labeled nuisance areas." Stressing that he was not questioning the desirability of controlling salt-marsh mosquitoes or the midges and flies which flourish in this habitat, he pointed out that remedies often have been more harmful than the original nuisance. He decried the sale of shellfish-producing tidal flats worth at least $4,200 per acre as "cheap fill" to dry up marshes—themselves worth at least $1,000 per acre before alteration. The "expendable" area where a few insects formerly flourished (together with birds, small mammals, reptiles, and fish) may then become "garbage dumps or parking lots for boats" or new housing developments which "invariably require additional expenditure of public funds. Furthermore, nutrient-rich cesspool effluents from houses along the shore usually seep into the adjoining waters and add to the pollution load."

In cases where the wetlands are not drained or filled, offending insects are destroyed with nonselective poisons. Both approaches have had a horrifying effect on an area which Taormina had been studying. Throughout Long Island, he reported, "dredges have been steadily chewing away" at valuable habitat, and the enthusiastic use of DDT and other pesticides has left alarmingly high concentrations of residues poisonous to man as well as animals. With dismay surfacing through statistical detachment, he wrote, "there is a surprisingly high amount of chlorinated hydrocarbons in the marine water of Long Island."

The acceleration of erosion is equally alarming:

Each dredging operation which removes the stable peaty salt marsh . . . increases the vulnerability of adjoining channels, shorelines, and manmade structures to damage from the storms as well as from accelerated tidal flow. . . . A classic result of such a disastrous dredging operation can be seen along the northwest edge of the Village of Belle Terre where it borders Port Jefferson Harbor. A once-beautiful and stable shore, protected by marsh and woodland, was undermined by sand and gravel dredging. The marshes and woodland are gone, the steep, bare banks are eroding rapidly and the people of Belle Terre are faced with a monumental, if not impossible, bank stabilization project.

A difficulty in digesting such reports is that hardly anyone can visualize the enormity of damage. Under normal circumstances "cesspool effluents" may go undetected, and it is an effort to believe the 1970 estimate of untreated sewage spewed into Chesapeake Bay—four hundred million gallons *daily*. That does not include the discharge of human waste from some six thousand merchant ships steaming in and out of Baltimore Harbor each year. The pollution promotes algal growth which is causing eutrophication of a bay that is also threatened by siltation and rising salinity resulting from harbor and canal dredging by the Army Corps of Engineers. Food chains are broken, life cycles altered or cut. The sewage has proved toxic to fish. If it poisons fish, will it not also kill birds? And people?

In answer, James B. Coulter, Maryland's Director of Natural Resources, announced an ambitious plan to end pollution and, within a decade, to eradicate eutrophication. The putrid inner harbor of Baltimore is to be transformed into a clean recreation area, and the state's revised water-quality standards were among the first to be approved under the federal Water Quality Act of 1965.

Maryland was also among the first states to supplement recently authorized federal grants for the improvement of sewage treatment. Bowie, Maryland, was the first town in the country to ban the sale of unreturnable bottles and cans. Other municipalities along the flyway were among the first to halt the sale of detergents containing phosphates which, like sewage, encourage eutrophication. Since urban regions are bound to suffer the most severe pollution, they are also the most prompt to act, and there is cause to hope that the seaboard states will lead an environmental revival.

Similarly, there is hope in the deluge of publicity that has accompanied the deluge of other pollutants in the brief time since research has proved the accuracy of the calamitous portents in Rachel Carson's *Silent Spring*. Dr. Lucille F. Stickel, a government-employed scientist at the Patuxent Wildlife Research Center in Laurel, Maryland, has warned that "each year for more than two decades, many millions of pounds of organo-chlorine pesticides have been applied to the land. DDT and Dieldrin now occur in all the major river basins of the United States."

A more recently exposed group of related poisons are the polychlorinated biphenyls, used in electrical equipment, paint, and other products. Their effects are similar to those of the previously suspected chlorinated hydrocarbon pesticides—a lethal thinning of fragile eggshells, interference with the reproductive system and embryonic development, and death for any animal that has ingested enough of the poison.

Still other poisons, such as strychnine, thallium, and cyanamide baits intended to kill predatory species (some of which are in dire need of protection rather than diminution) have killed significant numbers of migratory birds, fish, and other game animals. And the nitrates used in agricultural fertilizers have made drinking water dangerous to man in some areas.

In 1972, Secretary of the Interior Rogers C. B. Morton announced that the Fish and Wildlife Service would henceforth ban the use of poisons to kill predatory animals on federal lands. If the time comes when a ban can halt the poisoning of all wildlife, soil, and water, it will be one giant setback for industrial lobbyists, one small step for mankind.

Mercury has lately been ranked as potentially deadlier than DDT, though it is still used in the production of paints, plastics, fungicides, chlorine, caustic soda, paper, and other products, and it is still discharged into North American waters. Ducks taken in 1970 at Lake St. Clair, near Windsor, Ontario, appeared healthy but their tissues were found to contain more than twice the level of mercury that is legally deemed safe for consumption. That same year, hunters in some parts of Canada and northern New England were warned against eating woodcock for the same reason. (Since mercury, like the other persistent poisons, is concentrated and stored in the tissues, it is reasonable to suppose that birds considered toxic in Maine would be equally toxic when they reached Delaware or Georgia or Louisiana.) Subsequently, the Department of Agriculture prohibited the use of forty-eight different mercury compounds.

Therein lies one of the causes for optimism in this appalling recital of poisonings. Despite inevitable opposition from industrial and agricultural lobbyists, the federal government and a number of states and Canadian provinces have begun to curtail the indiscriminate use of harmful chemicals. Where bans have been ignored, polluters are being sued and some industrial plants have been closed until corrective measures can be completed. On the first day of 1971, New York State put into force the nation's most stringent controls on persistent pesticides. Ten of the most dangerous, including DDT and mercury compounds, were banned entirely, and severe restrictions were placed on the sale and use of about sixty others.

Moreover, the recognition of crisis has spurred much-needed research. Government and university scientists have succeeded in hatching endangered birds in incubators. A Cornell University project has also been successful in reproducing endangered species by means of artificial insemination. And, with governmental help, a young Cornell graduate student has been working toward his doctorate by collecting healthy osprey eggs and chicks at Chesapeake Bay for transfer to nests on Long Island Sound, where pesticides have rendered eggshells too fragile to support their own weight. Many species of birds will lay again if their eggs are taken, and will brood new ones put into their nests. This experiment, therefore, may prove valuable in the restoration of declining populations.

There seems to be no limit to the variety of pollutants which man, in his infinite ingenuity, can train upon the environment. In 1969, a barge ran aground off West Falmouth while en route to refuel a power plant on the Cape Cod Canal. The accident did not have the gargantuan dimensions of the infamous Santa Barbara spill. Only seven hundred tons of fuel, perhaps less, dribbled into Buzzards Bay. The oil degenerated into toxic compounds, though, as it always does, and the kill of marine animals was massive.

For two years, scientists of the Woods Hole Oceanographic Institution continued to take samples at the head of Wild Harbor River in West Falmouth, Massachusetts. They set up one of their sampling stations outside the expected spread of the oil, but within three weeks it had seeped past that point. In eight months, it had spread over several miles, affecting nearly six thousand acres of bay and marsh. After the beaches were cleaned, a casual observer might have thought the matter was ended unless he detected a foreign odor on the marshes. Actually, oil particles had broken up, sunk, and in some areas saturated the bottom to a depth of a foot and a half. The sunken slick began to recede in 1971, and marine life slowly started to recover. Only then did the resilient spartina grass begin to come back over twenty-two ruined acres of marsh. Investigators have predicted that the effects of the spill will probably be felt for many years.

But even with regard to oil pollution, there is cause for hope. Spill-proof methods of transportation are being sought, as are methods of removing spills so quickly that little or no damage can result. One research firm has developed an experimental microbial concoction that literally eats the oil—breaking it down into carbon dioxide, water, sugars, and proteins before harmlessly fading away for lack of anything else to eat.

When one considers that widgeon, canvasbacks, bluebills, and even a few black ducks are shot from sneakboats within sight of the residential buildings at Providence, Rhode Island, the quest for clean port waters assumes a terrible urgency. The officially listed species of endangered creatures rose from seventy-eight in 1968 to one hundred and two in 1971. Canvasback ducks, already "saved" more than once in recent history, dwindled by twenty-five percent between 1967 and 1971.

The Fish and Wildlife Service has assumed the task of accelerated research as well as continual surveys to determine conditions and prospects. The surveys are the world's most comprehensive and accurate. Airplane crews fly some eighty thousand miles in spring and summer to tally the birds on the breeding grounds in the northern United States and the most productive Canadian areas. Banding is conducted on a vast scale; wings sent in by hunters are examined for age and sex; the annual kill is carefully estimated; the wintering grounds are surveyed.

At Patuxent, game biologists have mastered the additional discipline of behavioral science, employing the discoveries of such pioneers as Konrad Lorenz. They place newly hatched black ducks in special nesting boxes to "imprint" them with the experience of using artificial, elevated, predator-proof nests. The ducklings are then reared very carefully under conditions simulating a natural environment so that they do not become tame and unable to survive in the wild. Attempts at the eradication of predators have failed throughout history to achieve their intended purpose, and now it has occurred to man that the raccoon and the crow and perhaps even the snapping turtle may have a place in the ecosystem—a right to live. Instead of annihilating these creatures, scientists are finding ways to shield the birds from excessive predation, and the outlook is excellent.

On the basis of releases to date, there is evidence that the imprinting technique instills in ducks a preference for nesting boxes, and these are being provided on the marshes. Evolution may actually be affected. A duck that began life in a nesting box is likely to seek the same kind of nest for its young; this clutch, in turn, will seek sheltered, elevated nests for its young, and the ground-nesting habits of an entire population may conceivably be altered as one generation after another assumes a new type of behavior in the wild.

There is also evidence that at least some species of wildfowl, if properly reared, will revert to a truly wild state, so that man's experiments will not weaken a wild strain of birds. The study has immense significance in the expansion of waterfowl populations. If a normally

183

In behavioral study at Patuxent research center, black ducks are "imprinted" with experience of nesting in predator-proof elevated boxes (below). Birds are carefully examined and banded (right). Observations of subsequently released ducks encourage speculation that nesting habits can be made safer, more productive.

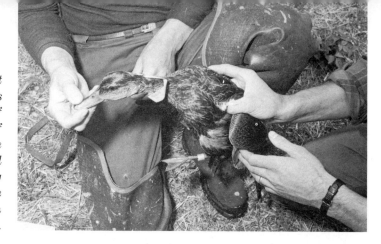

*On many small eastern ponds and marshes, hunter has better chance than heretofore of bagging limit of ducks.*

ground-nesting species like the black duck can be influenced to nest in a safer manner, the technique should succeed with other ground-nesting ducks, and perhaps geese, as well.

In scattered locations on the Atlantic Flyway, far different investigations are being pursued in an effort to conserve wildfowl. It has long been known that ducks and geese devour spent shot, especially the smaller sizes such as Number 6 and Number 7½, which do not sink into the mud as quickly as larger pellets. The shot is retained in the gizzard together with natural grit, slowly wearing away and producing lead poisoning which sometimes can be lethal. The major ammunition manufacturers have tried for years to find a nontoxic substitute for lead pellets. Almost everything imaginable has been tried, from plastic coatings to sintered iron. Some of the materials have been poor ballistically, others have been prohibitively expensive or severely damaging to gun barrels. While the government encouraged research, it has effectively discouraged genuine cooperation among five ammunition companies wrestling with the problem, because antitrust laws inveigh against any industrial closing of the ranks. Success may nonetheless be achieved in the not very distant future.

The latest proposed substitute for lead is a steel alloy which has already undergone some field testing. Initial results indicate that in a moderately large size (Number 4) the steel pellets are ballistically comparable with Number 6 lead shot and equally effective in bagging ducks at ranges up to at least forty yards. The steel appears to be harmless to gun barrels. If a way can be found to produce steel pellets of uniform shape and size as economically as lead pellets are produced, perhaps lead will some day be obsolete.

In terms of habitat, there have been still more encouraging developments. A recent study proved beyond doubt that channelization of nearby streams promotes excessive drainage of pothole areas. Channelization has been a pet activity of the Corps of Engineers for years, but the National Environmental Policy Act of 1969 now requires the Corps to submit a justification of environmental impact before inaugurating any new project. Channelization has become increasingly unpopular among conservationists and their legislators; henceforth it is likely to be employed with great restraint.

185

*Cover in some coastal areas still permits excitement of jump-shooting.*

In another study, conducted over a three-year period in a North Dakota breeding area, it was discovered that waterfowl nest in surprisingly large numbers and with surprising success along the right-of-way lands on the sides of highways, railways, utility lines, and pipelines. There are fifty million acres of these rights-of-way in the United States, and they can be managed effectively enough so that some states and perhaps the federal government will consider putting them to use.

In Georgia in 1968, public protest blocked the issuance of permits to a corporation that wished to mine phosphate on state-owned coastal marshes. Legislation was soon introduced to protect the coastal wetlands. The trend toward state protection of such areas soon spread along the flyway. The going was hard in states hungry for industry. In Delaware in 1961, it became known that an oil company held an option on thirty-five hundred acres of marsh in the famous Blackbird Hundred, frighteningly close to the Bombay Hook Wildlife Refuge and encompassing hunting grounds cherished since colonial times. When local citizens protested, the area was rezoned for heavy industry upon the basis that sixty percent of the area was "marsh and hence useless for any other use." The courts upheld the decision of the zoning commission. Fortunately, in this instance heavy industry restrained itself and will now be prevented from succumbing to any change of heart. More fortunately, this kind of pandering to industry will not be repeated in Delaware.

In 1971, Governor Russell Peterson proposed, fought for, and signed legislation banning any further heavy industry on the coast. The exceptionally strong law, first of its kind in the nation, even precludes offshore oil and coal transfer to terminals that had been planned.

This action required unusual courage on the part of the governor, legislators, and a number of individuals and groups who supported the proposal even though they could have gained great financial benefit from industrial development. The law was enacted during a period of recession, and in defiance of opposition from the United States Treasury and Department of Commerce. It has encouraged conservationists in other states to increase pressure for environmental protection.

In 1970, the Sierra Club reported that a New Jersey highway, in order to avoid disrupting a housing development, was consuming a large part of Great Piece Meadows, a resting and feeding site on the migratory lane; a second highway was cutting through the marsh of Troy Meadows; and a third was amputating a section of the Watchung Reservation. At about that time the Izaac Walton Club filed suit against a group of land developers and the Corps of Engineers to halt the dredging and filling of coastal wetlands around Cape May, New Jersey. A sprawling jet airport was proposed as an appropriate replacement for a tract of the beautiful Pine Barrens. Prospects grew steadily bleaker. But here, too, a new hope materialized.

Governor William T. Cahill signed legislation giving the state's Department of Environmental Protection power to regulate wetland development and to halt it if harm is being done to the ecological balance. Most of New Jersey's three hundred thousand acres of coastal marsh may yet be saved.

Elsewhere, too, the forces of destruction are at last being opposed. Early in 1972 Massachusetts authorized a major bond issue for the acquisition—salvation, that is—of endangered wetlands. The beginning of a trend can be discerned as other states take note of the need for action.

It is also true, however, that the battle for the Atlantic Flyway has barely been joined and vast areas of the seaboard face greater menaces than ever. The past two centuries of habitat destruction in the urbanized northeast have been well-publicized. The continuing destruction in the southeast, from the Atlantic to the Mississippi Flyway, has received less attention. Forestry records indicate that during the last decade three-hundred thousand acres per year of the finest mallard and wood-duck habitat have been destroyed in the big deltas of Arkansas, Louisiana, and Mississippi. The ecosystems under attack have chiefly been bottomlands of hardwoods interspersed with cypress and tupelo swamps, drained and channelized in agricultural and flood-control projects by the Corps of Engineers and the Soil Conservation Service. In a single decade in this area, some two million six hundred thousand acres of wildlife habitat have thus been sacrificed for supposed benefits which are now causing serious problems in soil and water management, and the battle must continue.

*Florida wildfowler continues to find birds converging on traditional*
*southern wintering spots, though wetland drainage poses unremitting threat.*

Since defeats always appear larger than the victories, overconfidence has not yet impeded the crusade. Despair is the more prevalent self-indulgence and an equally great violation of nature's economy. Fortunately, it has not yet infected the professional menders of ecological dikes. A short-term goal of the Interior Department has lately been the acquisition of some two and a half million more refuge acres. The Fish and Wildlife Service has struggled halfway toward that goal. Its spokesmen now express a fully justified mortification at soaring land costs and the paucity of appropriations, but conservationists have learned that rage overcomes despair and even apathy.

There are times, of course, when man's past indifferent wastefulness is mirrored (or mocked) by nature's own perversity. Two and a half miles off Maryland's Eastern Shore lie three remnants of land known as Poplar Island, Coaches Island, and Jefferson Island, middenlike jumbles of vines, fallen trees, bricks and oyster shells, vestiges of

THIS AUTUMN AND NEXT

*Snows, Canadas*
*and ducks are aloft,*
*men wait tensely*
*amid tall corn stalks,*
*straining to identify*
*legal species quickly*
*so that shots*
*need not be rushed—*
*and mallard*
*proves Chesapeake*
*Bay pup has learned*
*his lessons well.*

*Mallards, shovelers, and Canada geese are among many species certain to benefit from current management research. Pit-blind gunner in Atlantic Flyway cornfield can also be thankful for modern farming methods.*

There are annual visits from foreign delegations such as white-faced barnacle geese, but of greater importance is North America's ability to nurture more and more native wildfowl—mallard ducklings, Canada goslings, and other future wayfarers along all flyways.

building foundations, poison ivy, and forty-foot-high loblolly pines. The little archipelago sometimes goes by the collective name of Poplar Island, its title during the three centuries when it was a single land mass supporting farmers and watermen. It covered more than a thousand acres when Captain John Smith sighted it in 1608. Now the three islets amount to no more than a hundred and sixty-three acres and even the central portion, "Poplar Island proper," is no longer a single mass, having been channeled into wooded pieces by the erosion of wind and tide and storm. Some of its windward shoreline falls away at a rate of fourteen feet in an average year, and barely fifty-four acres remain.

A Virginia trader, William Claiborne, claimed the landfall in 1631 and named it Poplin's Island, after an associate. Early developments were inauspicious. Claiborne settled a family on it, whose members were massacred with inhospitable alacrity by Nanticoke Indians. A succession of owners corrupted the place name to Popeley's Island. For a while the Dutch governor of the future state of Delaware owned it, and later it was the property of Charles Carroll, who was among the signers of the Declaration of Independence. By that time it was called Poplar Island. During the nineteenth century and part of the twentieth, it supported a complacent little community of watermen, but wind and tide had pirated all but about five hundred acres in 1912. The land was soon serrated by channels.

Revenuers captured five bootleggers on Poplar in 1929. Then, in 1931, a group of Washington politicians bought a couple of separated pieces. On the larger one they built their spacious lodge and established the Jefferson Islands Club. Franklin Roosevelt visited it and partook of wild duck and terrapin; President Truman dined there, too. Eventually, the lodge burned and was replaced by a new club of Delaware sportsmen. Another fire, another club, and finally a considerable portion of the land was deeded to the Smithsonian in the hope of halting the continued erosion and saving the birds. Poplar had become the home of some thirty nesting pairs of ospreys, and hundreds of great blue herons were nesting in the pines. There were kingbirds, too, and crested flycatchers, house wrens and five-lined skinks and crustaceans and flowering lilies. It is an important breeding ground for herons and it is the site of the most successful osprey colony north of Florida. Those are the species that first present themselves to the casual birdwatcher. A wildfowler is likely to head for the lee of the island, and when the prow rounds the point his curiosity will be satisfied. It is a productive lee, a wintering lee for hundreds of ducks, geese, swans, grebes, mergansers, loons.

But the windward shore melts away and forty trees or more crash into the water each year. In a decade there will be only barren little hummocks. If erosion is not halted there will be no slatey herons or imperious ospreys. There will be no lee for the waterfowl.

Schemes for redemption have been proposed by Smithsonian investigators and others: landfills, a manmade beach with a protective revetment, jetties, a manmade lagoon, a marina, a wide extension of solid waste topped with landfill. The projected costs are enormous, and there are those who worry about setting a precedent that will lead to the pressure for landfills that has turned San Francisco Bay into a leased business district.

David Challinor, a Smithsonian scientist and a concerned conservationist, has heretically remarked that the wisest course may be to do no more than observe as nature takes its course. The entire vast Chesapeake Bay is a stripling among estuaries, created by glacial runoff perhaps eight thousand years ago and with a life expectancy of no more than an additional eight thousand years. "Siltation will probably fill it until a river meanders through Maryland to a delta near Norfolk, Virginia. . . . Parts of the Eastern Shore have lost two-thirds of an acre per mile *annually* for more than a century."

As the naturalist Hal Borland has often noted, it is sometimes best to leave nature alone and let it rebalance itself. *Of Men and Marshes,* a slim little volume by the late Paul Errington, tells how cautiously man ought to behave toward the fragile wetland ecosystems. And how he does behave, which has rarely been the same thing. Errington carefully chose his words, of which one was "arrogance." But he might well have argued with the use of the word "fragile" to describe the marshes, delicate as he knew they were. He also knew that man has trod upon those marshes for only a brief moment in this geologic epoch, and what he truly controls is only the length of time he struts and frets upon the stage.

"We hear," Errington said, "the chant of the juggernautists that nothing can withstand man." Later, in a less sardonic mood, he said, "No species has taught me more about parallels that I think man should be familiar with than has the muskrat, a living entity that, like man, has problems of living with what endowments it possesses, of meeting vicissitudes, and of getting along with its fellows."

He might have added that, like man, the muskrat has a proper place in nature, but nature, if pressed, has rather good prospects of managing without it. Perhaps only the sportsman and the conservationist understand the forbearance implicit in Errington's admonition to "reflect upon lessons and beauties that are not of human making."

# Chapter Fourteen

# EPILOGUE

The end of the hunting season is a reflective time for wildfowlers. The last patient interlude in a weatherbeaten blind is usually one of contentment even if the day is repellently bluebirdy and the horizon barren for long stretches. A hunter may whittle away hours remembering other days, imaginary horizons, favorite haunts that echo in the mind with the calls of elusive ducks and geese. There is also quiet pride that can be savored in memory. Pride in the stamina sometimes required for the enjoyment—genuine, wondering, thanksgiving enjoyment—of the worst weather nature can concoct, the steadfastly nasty kind that is often the best wildfowling weather. If the hunter is disposed to self-analysis, perhaps he grins at the recognition of something more complex than pride: the joy of getting along comfortably with nature, accepting what is offered, passing from the modern chaos of society to the experience of a still primeval marsh. And the joy of ever-deepening knowledge. Transcending even the gratification of the difficult shot well made is the calm yet exhilarating pleasure of initiation into the mysteries of migrations and many other behavior patterns of wildlife, the miraculous interrelationships of all living things in an ecosystem.

A season's last day can also invoke uncommonly placid reflections on the transience of life. A wildfowler who absently ruffles the down of a black duck, its back lying heavily on his palm, its head hanging down and one wing askew, may be silently accepting his own mortality.

The marsh seethes with life, death, life-giving death. Man himself is prey as well as predator. In a different but relevant context, Victor Scheffer remarked that the enemies of the great carnivores are no less menacing for being small and insidious. "Erosive, unimpressive costs of living," he called them in *The Year of the Whale:* "storms and droughts, fires and floods, days of starvation . . . scratches, infections, tumors . . . poisonous foods . . . all the thousand natural shocks that flesh is heir to. Death comes . . . slowly, not as the quick extinction of a rabbit surprised at his nest by a fox." As death stalks the whale or the lion, so it stalks a man. And, the hunter might add, so it stalks those birds that are not surprised in the sky by a quick shot or at the nest by some other predator.

But death is the progenitor and food of life. Black ducks nesting on a tidal marsh may raise their young in a welter of riches—insects and plants, mussels and snails; each morsel nibbled is a morsel killed. Dead, too, is the broken shell left by the hatchling. It decomposes amid the snail shells and the droppings of feeding birds, seed husks and husks of moulted insects, the bones and flesh, feathers or fur or scales of other creatures, dead but not obliterated, the decaying cordgrass or salt hay, or any of the other spartinas, or any other marsh vegetation, all blending to form the compost on which depends the life of the marsh. Coarse and humble cordgrass, of mediocre value as food for birds, has inestimable value as food for the procession of life that ultimately feeds those birds.

When the peaty salt marsh quakes underfoot, it throbs with life born of death. So, too, does the death of a barrier dune or an entire tidal flat signal the birth of another. Though five thousand years may be required for the sea to rise thirty feet, the nibbling is relentless. Erosion carries away sand and detritus, and a state somewhere on the Atlantic Flyway loses fifteen square miles or more of coastal marsh in a scant century. But of course nothing in nature is actually lost. Elsewhere on the flyway the sand is deposited again. Poplar Island subsides into Chesapeake Bay, Hog Island is devoured by the waters off Virginia, but somewhere, slowly, new dunes rise. Lagoons form. Marsh grasses sprout.

Man's erosive hunger does not, however, invariably build as it destroys. The wildfowler seeks comfort in the reflection that even man is capable of learning. Sometimes the learning is dangerously slow. All but a couple of million acres of the East Coast's salt marshes have been despoiled. When so expressed, the situation appears irreversibly disastrous, yet two million acres, together with inland havens, have permitted the creatures of the Atlantic Flyway to survive. Many species, in fact, are flourishing. Most scientists—trained skeptics not often prey to foolish optimism—agree that the remaining marshes are themselves surviving, and can continue to if modern enlightenment does not dim.

States now vie with one another for the role of legislative exemplar, and bond issues proliferate to finance the acquisition of wetlands, but single remedies will not be decisive. In some areas, perhaps an expansion of the government's Water Bank Program will encourage farmers to maintain wetlands which might otherwise endanger their own economic survival. In those and other areas, the growth of outdoor recreation as an industry may spur restoration and preservation, or it may continue to replace swamps with motorboat marinas. In most areas, federal and state agencies for environmental protection will reduce the profit and acceptance of destruction. In all areas, wildfowlers and their fellow conservationists inevitably will lead an aroused but irresolute citizenry toward the perpetuation of their own and other species.

Survival, endurance, indomitability. They are attributes of migratory birds and all wild creatures, of the wetlands and all of nature. In man, a unique fusion of those and other attributes is called conscience by those who acknowledge that morality is chiefly a concern for the survival of one's own species. There are signs that man's conscience has at last awakened to the alarums of visionaries like William Beebe, who once wrote that "the beauty and genius of a work of art may be reconceived though its first material expression be destroyed; a vanished harmony may yet again inspire the composer; but when the last individual of a race of living things breathes no more, another heaven and another earth must pass before such a one can be again."

When the wildfowler quits his blind for the season, days have begun to lengthen and some of the birds in southern marshes are growing restless, already yearning for the north. The hunter, watching a little wedge of ducks melt into a northern horizon, wishes them farewell. He visualizes their arrival at the nesting grounds and thinks about life cycles, sowing and harvesting, eternal renewal, the complex simplicity of his planet. A season's end, he observes with equanimity, is also a season's beginning.

# BIBLIOGRAPHY

Audubon, John James  *The Birds of America,* first reprint of Audubon's 1840-44 plates and legends, Macmillan, 1937; and Dover Publications reprints of texts ("Ornithological Biography") published with 1840-44 *Birds of America* editions, 7 vol.

Barber, Joel D.  *Wild Fowl Decoys,* Windward House, 1934; and Dover Publications (with biographical preface and illustrative appendix), 1954.

Bellrose, Frank C.  *Migrational Behavior of Mallards and Black Ducks as Determined from Banding* (by Frank C. Bellrose and Robert D. Crompton), Illinois Natural History Survey Bulletin, Vol. 30, Article 3, Natural History Survey Div., 1970.

Bent, Arthur Cleveland  *Life Histories of North American Wild Fowl,* Smithsonian Institution, National Museum Bulletins 126 and 130, 1923 and 1925; reprint editions, Dover Publications, 1962.

Buckingham, Nash  "The Prodigal Years," article in *The American Sportsman,* Vol. 2, No. 1, Winter 1969, The Ridge Press, Inc.

Camp, Raymond R.  "Waterfowl of the Outer Banks," article in *The American Gun,* Vol. 1, No. 1, Winter 1961, Madison Books, Inc.
——*The Hunter's Encyclopedia,* The Stackpole Co., third edition, 1966, and revised third edition, 1972.

Carson, Rachel  *Silent Spring,* Houghton Mifflin, 1962.

Clark, Roland  *Gunner's Dawn,* Derrydale Press, 1937.
*Pot Luck,* A. S. Barnes, 1945.

Cleveland, Grover  *Fishing and Shooting Sketches,* The Outing Publishing Co., 1906; reprint edition by Abercrombie & Fitch, 1966.

Connett, Eugene V., III  *Duck Shooting Along the Atlantic Tidewater,* William Morrow & Co., 1947.

Coues, Elliot  *Key to North American Birds,* 5th edition (revised), 2 vol., Estes Co., 1903.

Crowell, A. Elmer  "Cape Cod Memories," reminiscences in *Duck Shooting Along the Atlantic Tidewater,* by Eugene V. Connett, III, William Morrow & Co., 1947.

Delacour, Jean  *The Waterfowl of the World*, 4 vol., Country Life, 1954-64.

Elman, Robert  *The Great Guns* (by Harold L. Peterson and Robert Elman), Ridge Press/Grosset & Dunlap, 1971.
——— *The Great American Shooting Prints*, Ridge Press/Alfred A. Knopf, Inc., 1972.

Errington, Paul  *Of Men and Marshes,* Iowa State University Press, 1957.

Falk, John R.  *The Practical Hunter's Dog Book*, Winchester Press, 1971.

Forbush, Edward Howe  *Birds of Massachusetts and Other New England States,* Mass. Dept. of Agriculture (printed by Berwick & Smith), 2 vol., 1925-27.

Forester, Frank  See Herbert, Henry William.

Gohdes, Clarence  *Hunting in the Old South,* Louisiana State University Press, 1967.

Greener, W. W.  *The Gun and Its Development,* 1881; reprint of ninth edition, Bonanza Books, 1968.

Grinnell, George Bird  *American Duck Shooting,* Forest and Stream Publishing Co., 1901.

Heilner, Van Campen  *A book on Duck Shooting,* Alfred A. Knopf, Inc., 1939.
"Brant: Harvest on the Marsh," article in *The American Gun,* Vol. 1, No. 3, Summer 1961, Madison Books, Inc.

Herbert, Henry William ("Frank Forester")  *The Complete Manual for Young Sportsmen,* Stringer & Townsend, 1856.

Hornaday, William Temple  *Our Vanishing Wild Life,* Charles Scribner's Sons, 1913.

Johnsgard, Paul A.  *Waterfowl, Their Biology and Natural History,* University of Nebraska Press, 1968.

Koller, Larry  *The Treasury of Hunting,* Ridge Press/Odyssey Press, 1965.

Kortright, Francis H.  *The Ducks, Geese and Swans of North America,* American Wildlife Institute, 1942; reprint edition by The Stackpole Co. and Wildlife Management Institute, 1967.

Lewis, Elisha Jarrett  *The American Sportsman,* Lippincott, Grambo & Co., 1855.

Lincoln, Frederick C.  *The Waterfowl Flyways of North America,* U.S. Dept. of Agriculture Circular 342, 1935.
——— "Migration Routes and Flyways," article in *The Ducks, Geese and Swans of North America,* by Francis H. Kortright, American Wildlife Institute, 1942.

Linduska, Joseph P.  *Waterfowl Tomorrow,* Fish and Wildlife Service; U.S. Government Printing Office, 1964.

Lunt, Dudley Cammett   *Taylors Gut in The Delaware State*, Alfred A. Knopf, Inc., 1968.
—— "Voices over Gardner's Marsh," article in *The American Sportsman*, Vol. 3, No. 1, Winter 1970, The Ridge Press.

Mackey, William J., Jr.   *American Bird Decoys*, Dutton, 1965.
—— Foreword to *The Classic Decoy Series*, by Milton C. Weiler, Winchester Press, 1969.
—— Text for *Classic Shorebird Decoys*, by Milton C. Weiler, Winchester Press, 1971.

Madson, John   *The Mallard*, Olin Mathieson Conservation Dept., 1960.

Martin, Alexander C.   *American Wildlife & Plants* (by Alexander C. Martin, Herbert S. Zim, Arnold L. Nelson), McGraw-Hill, 1951; reprint edition, Dover Publications, 1961.

Matthiessen, Peter   *Wildlife in America*, Viking Press, 1959.

Megargee, Harry   "Barnegat Sneakbox," article in "Sportsmen Afloat" column, *Field & Stream*, December, 1942.
—— Reprinted construction drawings and description of duck boat in "Barnegat Bay Sneakbox," by F. M. Paulson, *Field & Stream*, October, 1971.

Neely, William W.   *Wild Ducks on Farmland in the South* (by William W. Neely and Verne E. Davison), U.S. Dept. of Agriculture Farmers' Bulletin 2218, revised edition, U.S. Government Printing Office, 1971.

Nevins, Allan   *Grover Cleveland: A Study in Courage*, Dodd, 1933.

Osborn, Henry Fairfield   Preface to *Our Vanishing Wild Life*, by William T. Hornaday, Charles Scribner's Sons, 1913.

Parker, Eric   *Colonel Hawker's Shooting Diaries*, Derrydale Press, 1931.

Peterson, Harold L.   *The Treasury of the Gun*, Ridge Press/Golden Press, 1962. *Pageant of the Gun*, Doubleday & Co., 1967.
—— *The Great Guns* (by Harold L. Peterson and Robert Elman), Ridge Press/Grosset & Dunlap, 1971.

Phillips, John Charles   *A Natural History of the Ducks*, 4 vol., Houghton Mifflin Co., 1922-26.

Roth, Charles B.   "Fred Kimble: Champion Duck Shooter and a Discoverer of the Chokebore Shotgun," article in *The American Gun*, Vol. 1, No. 2, Spring 1961, Madison Books, Inc.

Rue, Leonard Lee, III   *New Jersey Out-of-Doors: A History of its Flora and Fauna*, Hicks Printing Co., 1964.

Socrates   Passages in Plato's "Dialogues," *The Works of Plato*, selected and edited by Irwin Edman, Modern Library, 1956.

Terres, John K.  *Flashing Wings: The Drama of Bird Flight*, Doubleday & Co., 1968.

Thorpe, Thomas Bangs  "Woodcock Fire-Hunting," article in *The Spirit of the Times*, May 1, 1841; reprinted in *The Hive of "The Bee-Hunter,"* D. Appleton & Co., 1854, and in *Hunting in the Old South*, edited by Clarence Gohdes, Louisiana State University Press, 1967.

Townsend, George Alfred  *Tales of the Chesapeake*, 1880; reprint edition, Tidewater Publishers, 1968.

Wallace, William N.  *The Macmillan Book of Boating*, Ridge Press/Macmillan, 1964.

Walsh, Harry M.  *The Outlaw Gunner*, Tidewater Publishers, 1971.

Walsh, Roy E.  *Gunning the Chesapeake: Duck and Goose Shooting on the Eastern Shore*, Tidewater Publishers, 1960.

Webster, David S., and Kehoe, William  *Decoys at Shelburne Museum*, Museum Pamphlet Series, No. 6, The Shelburne Museum, revised edition, 1971.

Weiler, Milton C.  *The Classic Decoy Series*, Winchester Press, 1969.
——*Classic Shorebird Decoys*, Winchester Press, 1971.

Whitehead, Charles Edward  "A Duck Hunt in Florida" and other "Camp-Fire Stories," reprinted from *The Spirit of the Times* in *Wild Sports in the South; or, The Camp-Fires of the Everglades*, Derby & Jackson, 1860; and in *Hunting in the Old South*, edited by Clarence Gohdes, Louisiana State University Press, 1967.

Williams, C. S.  *Honker*, D. Van Nostrand Co., Inc., 1967.

Wilson, Alexander  *American Ornithology*, reprint of 1840 edition, Arno, 1970.

Wilstach, Paul  *Tidewater Maryland*, Tidewater Publishers, 1969.

Zern, Ed  Text for *The Classic Decoy Series*, by Milton C. Weiler, Winchester Press, 1969.

# SOURCES OF BLACK AND WHITE ART AND SUPPLEMENTARY PHOTOS

Abercrombie & Fitch/University Microfilms (illustrations from Grover Cleveland's *Fishing and Shooting Sketches*) : pages 100, 101, 158, 159.

The Brooklyn Museum : page 132 (top).

Canadian Government Travel Bureau : page 41.

Russ Carpenter : page 24.

Culver Pictures : pages 66, 67.

Charley Dickey : pages 18, 27, 49, 54, 57, 87, 103, 104, 105, 139, 185, 186.

Doubleday & Company (illustrations from Harold L. Peterson's *Pageant of the Gun*) : **page 115**.

Ducks Unlimited : page 37.

Elman Pictorial Collection : pages 6 (bottom), 17, 40, 88, 102, 108, 131 (top).

*Field & Stream:* page 127.

Florida News Bureau : page 188 (photo by Johnny Johnson).

Florida State University Library : pages 122, 123.

Gaines Dog Research Center : page 85 (top).

Gould Collection, University of Michigan : page 156.

Gun Digest Company : page 113.

Historical Society of Pennsylvania : page 83.

Nicholas Karas : page 63.

Kennedy Galleries : pages 131 (right), 134.

Library of Congress : pages 45, 70, 72, 78, 129, 131 (bottom), 132 (center and bottom), 141, 155 (bottom), 160 (left), 167 (bottom).

Louisiana State University Press (illustrations from Clarence Gohdes' *Hunting in the Old South*) : pages 153, 154.

Karl H. Maslowski : page 28.

Memorial Art Gallery of the University of Rochester : page 69.

Crosby Milliman Collection : page 107 (botttom).

Museum of Fine Arts, Boston : page 89.

National Audubon Society : pages 75 (left and right), 76.

New York Public Library Picture Collection: pages 157, 163.
North Carolina Wildlife Resources Commission: page 9
(photo by Joel Arrington).
Old Colony Historical Society: page 107 (top).
Old Print Shop: pages 111, 117.
Ontario Department of Travel & Publicity: page 7.
Harry T. Peters Collection, Museum of the City of New York: page 73.
Remington Art Memorial Museum: page 155 (top).
Leonard Lee Rue:
pages 2, 5, 13, 14, 38, 43, 52, 80, 128, 137, 142, 144, 145, 151, 167 (top), 178.
Saskatchewan Government Photographic Services: page 176.
Possessions of Victor D. Spark: pages 84, 160 (right).
Tennessee Game & Fish Commission: page 51 (photo by Charles Jackson).
United States Fish & Wildlife Service:
maps—pages 10, 11, 171, 172; photos—pages 4 (by W. F. Kubichek),
21 (by E. R. Quortrup), 44 (bottom) (by Julian Howard),
74 (by L. H. Walkinshaw), 119 (by R. G. Schmidt),
165 and 174 (by P. J. Van Huizen), 184 (by Luther Goldman).
University of Oregon Art Museum: page 62.
Milton C. Weiler: page 96.
Western History Research Center, University of Wyoming: page 6 (top).
Winchester Gun Museum: page 109.
Winchester News Bureau: pages 36, 75 (center), 79, 85 (bottom).
Yale University Art Gallery: page 44 (top).

# SUPPLEMENTARY ILLUSTRATIONS
# IN COLOR PORTFOLIOS

Portfolio 1, ORIGINS AND DEPARTURES: last page (top) by Robert Elman.
Portfolio 2, COASTAL INTERLUDES: sixth page (left) by Roy Attaway.
Portfolio 4, SHARERS OF THE WETLANDS: third page (bottom) by
Robert Elman.
Portfolio 6, AUTUMNS OF THE PAST: first page (bottom) from author's
print collection; second page, courtesy of Crossroads of Sport;
fifth page (right), courtesy of the Chicago Historical Society;
last page (bottom), courtesy of the Buffalo and Erie County
Historical Society.
Portfolio 7, THIS AUTUMN AND NEXT: sixth page (bottom) by Robert Elman.